WELCOME TO THE UNIVERSE

THE PROBLEM BOOK

PRINCETON
UNIVERSITY
PRESS

PRINCETON
& OXFORD

WELCOME TO THE UNIVERSE

THE PROBLEM BOOK

NEIL deGRASSE TYSON

MICHAEL A. STRAUSS

J. RICHARD GOTT

Published by Princeton University Press,
41 William Street, Princeton, New Jersey 08540

In the United Kingdom: Princeton University Press,
6 Oxford Street, Woodstock, Oxfordshire OX20 1TR

press.princeton.edu

Library of Congress Cataloging-in-Publication Data

Tyson, Neil deGrasse, author. | Strauss, Michael Abram, author. | Gott, J. Richard, author.
Welcome to the universe : the problem book / Neil deGrasse Tyson,
    Michael A. Strauss and J. Richard Gott.
Princeton, New Jersey ; Oxford : Princeton University Press, [2017]
LCCN 2017019526 | ISBN 9780691177809 (hardback ; alk. paper) | ISBN 0691177805
    (hardback ; alk. paper) | ISBN 9780691177816 (pbk. ; alk. paper)
    | ISBN 0691177813 (pbk. ; alk. paper)
LCSH: Cosmology—Problems, exercises, etc. | Stars—Problems, exercises, etc.
    | Relativity (Physics)—Problems, exercises, etc.
LCC QB981 .T975 2017 | DDC 523.1076—dc23 LC record available at
    https://lccn.loc.gov/2017019526

This book has been composed in Computer Modern and Trade Gothic LT Std

Printed on acid-free paper.

Printed in the United States of America

10 9 8 7 6 5 4 3

CONTENTS

Preface   xvii

Math Tips   xxi

PART I. STARS, PLANETS, AND LIFE   1

1 | THE SIZE AND SCALE OF THE UNIVERSE   3

1 Scientific notation review   3
Writing numbers in scientific notation.

2 How long is a year?   3
Calculating the number of seconds in a year.

3 How fast does light travel?   3
Calculating the number of kilometers in a light-year.

4 Arcseconds in a radian   3
Calculating the number of arcseconds in a radian, a number used
whenever applying the small-angle formula.

5 How far is a parsec?   3
Converting from parsecs to light-years and astronomical units.

6 Looking out in space and back in time   4
Exploring the relationship between distance and time
when traveling at the speed of light.

7 Looking at Neptune   4
The time for light to travel from Earth to the planet Neptune depends on where it and we are in our respective orbits.

8 Far, far away; long, long ago   5
There is an intrinsic time delay in communicating with spacecraft elsewhere in the solar system or elsewhere in the Milky Way galaxy.

9 Interstellar travel   6
Calculating how long it takes to travel various distances at various speeds.

10 Traveling to the stars   6
Calculating how long it would take to travel to the nearest stars.

11 Earth's atmosphere   7
Calculating the mass of the air in Earth's atmosphere, and comparing it with the mass of the oceans.

2 | FROM THE DAY AND NIGHT SKY TO PLANETARY ORBITS   8

12 Movements of the Sun, Moon, and stars   8
Exploring when and where one can see various celestial bodies.

13 Looking at the Moon   8
There is a lot you can infer by just looking at the Moon!

14 Rising and setting   9
Questions about when various celestial bodies rise and set.

15 Objects in the sky   9
More questions about what you can learn by looking at objects in the sky.

16 Aristarchus and the Moon   10
Determining the relative distance to the Moon and the Sun using high-school geometry.

17 The distance to Mars   11
Using parallax to determine how far away Mars is.

18 The distance to the Moon   11
Using parallax to determine how far away the Moon is.

19 Masses and densities in the solar system   11
Calculating the density of the Sun and of the solar system.

3 | NEWTON'S LAWS   13

20 Forces on a book   13
Using Newton's laws to understand the forces on a book resting on a table.

21 Going ballistic   13
Calculating the speed of a satellite in low Earth orbit.

22 Escaping Earth's gravity?   14
Calculating the distance at which the gravitational force from Earth
and the Moon are equal.

23 Geosynchronous orbits   14
Calculating the radius of the orbit around Earth that is synchronized
with Earth's rotation.

24 Centripetal acceleration and kinetic energy in Earth orbit 14
Calculating the damage done by a collision with space debris.

**25 Centripetal acceleration of the Moon and the
law of universal gravitation**   15
Comparing the acceleration of the Moon in its orbit to that of a dropped
apple at Earth's surface.

26 Kepler at Jupiter   16
Applying Kepler's laws to the orbits of Jupiter's moons.

27 Neptune and Pluto   17
Calculating the relationship of the orbits of Neptune and Pluto.

28 Is there an asteroid with our name on it?   17
How to deflect an asteroid that is on a collision course with Earth.

29 Halley's comet and the limits of Kepler's third law   18
Applying Kepler's third law to the orbit of Halley's comet.

30 You cannot touch without being touched   19
The motion of the Sun due to the gravitational pull of Jupiter.

31 Aristotle and Copernicus   19
An essay about ancient and modern views of the heavens.

4–6 | HOW STARS RADIATE ENERGY   20

32 Distant supernovae   20
Using the inverse square law relating brightness and luminosity.

33 Spacecraft solar power   20
Calculating how much power solar panels on a spacecraft can generate.

34 You glow!   21
Calculating how much blackbody radiation our bodies give off.

35 Tiny angles   21
Understanding the relationship between motions in space and in the
plane of the sky.

36 Thinking about parallax   22
How nearby stars appear to move in the sky relative to more distant stars,
due to the Earth's motion around the Sun.

37 Really small angles and distant stars   22
The Gaia spacecraft's ability to measure parallax of distant stars.

38 Brightness, distance, and luminosity   23
Exploring the relationship between brightness and luminosity of
various stars.

39 Comparing stars   23
Relating the luminosity, radius, surface temperature, and distance of stars.

40 Hot and radiant   24
Exploring the relation between the properties of stars radiating
as blackbodies.

41 A white dwarf star   24
Calculating the distance and size of a white dwarf star.

42 Orbiting a white dwarf   24
Using Kepler's third law to determine the orbit around a white dwarf star.

43 Hydrogen absorbs   25
Using the spectrum of an F star to understand the energy levels of a
hydrogen atom. A challenge problem.

7–8 | THE LIVES AND DEATHS OF STARS   27

44 The shining Sun   27
Calculating the rate at which hydrogen fuses to helium in the core
of the Sun.

45 Thermonuclear fusion and the Heisenberg uncertainty principle   27

Using quantum mechanics to determine the conditions under which thermonuclear fusion can take place in the core of a star. A challenge problem.

46 Properties of white dwarfs   29

Using direct observations of a white dwarf to determine its radius and density. A challenge problem.

47 Squeezing into a white dwarf   30

Determining how far apart the nuclei in a white dwarf star are.

48 Flashing in the night   30

Determining whether the gravity of a pulsar is adequate to hold it together as it spins.

49 Life on a neutron star   31

Calculating the effects of the extreme gravity of a neutron star.

50 Distance to a supernova   31

Watching a supernova remnant expand, and using this to determine how far away it is.

51 Supernovae are energetic!   32

Putting the luminosity of a supernova in context.

52 Supernovae are dangerous!   33

What would happen if a supernova were to explode within a few hundred light-years of Earth?

53 Neutrinos coursing through us   33

Calculating the flux and detectability of neutrinos emitted during a supernova explosion.

54 A really big explosion   34

Calculating the energy associated with a gamma-ray burst.

55 Kaboom!   36

Calculating the properties of one of the most powerful gamma-ray bursts ever seen.

56 Compact star   36

Calculating the distance between nuclei in a neutron star.

57 Orbiting a neutron star   37

Applying Kepler's third law for an orbit around a neutron star.

58 The Hertzsprung-Russell diagram   37
An essay about the relationship between surface temperature and
luminosity of stars.

9 | WHY PLUTO IS NOT A PLANET   38

59 A rival to Pluto?   38
Calculating the properties of a large Kuiper Belt Object in the outer solar
system, working directly from observations. A challenge problem.

60 Another Pluto rival   41
Exploring the properties of another large body in the outer solar system.

61 Effects of a planet on its parent star   43
Using observations of the motion of a star under the gravitational influence
of an orbiting planet to infer the properties of that planet.

62 Catastrophic asteroid impacts   44
How the impact of an asteroid on the early Earth may have evaporated
the oceans.

63 Tearing up planets   45
Calculating the tidal force of a planet on an orbiting moon.
A challenge problem.

10 | THE SEARCH FOR LIFE IN THE GALAXY   47

64 Planetary orbits and temperatures   47
Calculating the orbits and equilibrium temperatures of planets
orbiting other stars.

65 Water on other planets?   47
Determining whether liquid water can exist on the surface of planets
orbiting other stars.

66 Oceans in the solar system   48
Exploring the properties of oceans on Earth, Mars and Europa.

67 Could photosynthetic life survive in Europa's ocean?   49
Determining how much life the sunlight that impinges on Europa could support.

68 An essay on liquid water   50
An essay describing the conditions under which liquid water,
necessary for life as we know it, exists.

PART II. GALAXIES   51

**11–13 | THE MILKY WAY AND THE
UNIVERSE OF GALAXIES**   53

69 How many stars are there?   53
Calculating the number of stars in the observable universe.

70 The distance between stars   54
Putting the distance between stars into perspective.

71 The emptiness of space   54
Calculating the density of the Milky Way and of the universe as a whole.

72 Squeezing the Milky Way   54
What would happen if you brought all the stars in the Milky Way into
one big ball?

73 A star is born   55
How much interstellar gas do you need to bring together to make a star?

74 A massive black hole in the center of the Milky Way   55
Calculating the mass of the black hole at the center of our Galaxy,
working directly from observations.

75 Supernovae and the Galaxy   55
How many supernovae are needed to create the heavy elements in the
Milky Way?

76 Dark matter halos   56
Calculating the mass of the Milky Way from its observed rotation.

77 Orbiting Galaxy   57
The orbit of the Large Magellanic Cloud around the Milky Way, and what
it says about the mass of our Galaxy.

78 Detecting dark matter   57
Calculating how many dark matter particles there are all around us, and
how we plan to detect them. A challenge problem.

79 Rotating galaxies   60
Determining whether we can see the rotation of a galaxy on the sky.

80 Measuring the distance to a rotating galaxy   60
Using the apparent motion of stars in a galaxy in the plane of the sky
and along the line of sight to determine its distance.

14 | THE EXPANSION OF THE UNIVERSE 61

81 The Hubble Constant 61
Measuring the expansion rate of the universe from the measured properties
of galaxies.

82 Which expands faster: The universe or
the Atlantic Ocean? 63
The answer may surprise you.

83 The third dimension in astronomy 63
An essay about how we measure distances in the universe.

84 Will the universe expand forever? 63
The relationship between the density of the universe and its future fate.
A challenge problem.

85 The motion of the Local Group through space 64
Calculating the gravitational pull from the Virgo galaxy supercluster on our
Local Group of galaxies. A challenge problem.

15–16 | THE EARLY UNIVERSE AND QUASARS 67

86 Neutrinos in the early universe 67
Calculating just how numerous the neutrinos produced soon after the
Big Bang are.

87 No center to the universe 68
A brief essay explaining why the expanding universe has no center.

88 Luminous quasars 68
Calculating the properties of quasars, and the supermassive black holes
that power them.

89 The origin of the elements 68
An essay describing how different elements are formed in the universe.

PART III. EINSTEIN AND THE UNIVERSE 69

17–18 | EINSTEIN'S ROAD TO SPECIAL RELATIVITY 71

90 Lorentz factor 71
Exploring the special relativistic relation between lengths as seen in
different reference frames.

91 Speedy muons   72
How special relativity is important in understanding the formation and detection of muons created in the upper atmosphere.

92 Energetic cosmic rays   73
Determining relativistic effects for one of the highest-energy particles ever seen.

93 The *Titanic* is moving   73
Playing relativistic games with the great ship *Titanic*.

94 Aging astronaut   74
Understanding how the relativistic effects of moving an astronaut at close to the speed of light.

95 Reunions   74
How two friends can differ on the passage of time.

96 Traveling to another star   74
Calculating how time ticks slower for an astronaut traveling at close to the speed of light.

97 Clocks on Earth are slow   74
Calculating the difference between a clock in orbit around the Sun and one standing still.

98 Antimatter!   74
Should you run if trucks made of matter and antimatter collide with one another?

99 Energy in a glass of water   75
Calculating how much energy could be extracted from the fusion of the hydrogen in a glass of water.

100 Motion through spacetime   75
Drawing the path of the Earth's orbit around the Sun in spacetime.

101 Can you go faster than the speed of light?   75
Why the postulates of special relativity do not allow travel faster than the speed of light.

102 Short questions in special relativity   76
Quick questions which can be answered in a few sentences.

19 | EINSTEIN'S GENERAL THEORY OF RELATIVITY 77

103 Tin Can Land 77
Exploring the nature of geodesics on a familiar two-dimensional surface.

104 Negative mass 78
Would a dropped ball of negative mass fall down?

105 Aging in orbit 79
Exploring special and general relativistic effects on your clock while in orbit.
A challenge problem.

106 Short questions in general relativity 80
Quick questions that can be answered in a few sentences.

20 | BLACK HOLES 82

107 A black hole at the center of the Milky Way 82
Calculating the properties of the supermassive black hole at the center of
our Galaxy.

108 Quick questions about black holes 82
Short questions that can be answered in a few sentences.

109 Big black holes 83
Exploring the properties of the biggest black holes in the universe.

110 A Hitchhiker's challenge 83
The Hitchhiker's Guide to the Galaxy inspires a problem on black holes.
A challenge problem.

111 Colliding black holes! 84
Measurements of gravitational waves from a pair of merging black holes
allows us to determine their properties.

112 Extracting energy from a pair of black holes 85
Using ideas from Stephen Hawking to determine how much energy can be
released when black holes collide.

**21 | COSMIC STRINGS, WORMHOLES,
AND TIME TRAVEL** 87

113 Quick questions about time travel 87
Short questions that can be answered in a few sentences.

114 Time travel tennis   87
Playing a tennis game with yourself with the help of time travel.
A challenge problem.

115 Science fiction   89
Writing a science fiction story that uses concepts from astrophysics:
the challenge is to make it as scientifically realistic as possible.

**22 | THE SHAPE OF THE UNIVERSE
AND THE BIG BANG**   91

116 Mapping the universe   91
Ranking the distance of various astronomical objects from the Earth.

117 Gnomonic projections   91
Exploring the geometry of an unusual mapping of the night sky onto a flat
piece of paper.

118 Doctor Who in Flatland   95
Using concepts from general relativity to understand the nature of
Dr. Who's Tardis.

119 Quick questions about the shape of the universe   96
Short questions that can be answered in a few sentences.

**23 | INFLATION AND RECENT
DEVELOPMENTS IN COSMOLOGY**   97

120 The earliest possible time   97
Calculating it using both general relativity and quantum mechanics.

121 The worst approximation in all of physics   98
Can the Planck density give us a reasonable estimate for the density of
dark energy? Hint: no.

122 Not a blunder after all?   99
Describing the relationship between Einstein's desire for a static universe
and the accelerated expansion we now observe.

123 The Big Bang   99
An essay describing the empirical evidence that the universe started in
a Big Bang.

24 | OUR FUTURE IN THE UNIVERSE — 100

124 Getting to Mars — 100
Calculating the most efficient orbit to get from Earth to Mars.

125 Interstellar travel: Solar sails — 101
Using the pressure of light from the Sun to propel a spacecraft for interstellar travel.

126 Copernican arguments — 102
Applied to time.

127 Copernicus in action — 102
An essay about Copernican arguments in our understanding of the structure of the universe and our place in it.

128 Quick questions for our future in the universe — 103
Short questions that can be answered in a few sentences.

129 Directed panspermia — 103
Exploring how humankind could colonize the Milky Way with robotic probes.

Useful Numbers and Equations — 107

Solutions — 113

PREFACE

Our book, *Welcome to the Universe: An Astrophysical Tour*, grew out of an introductory survey course in astrophysics that Neil deGrasse Tyson, J. Richard Gott, and I team-taught for a number of years at Princeton University. The course is meant for nonscience majors, but we designed it to be a quantitative course. With mathematics no more complicated than high-school algebra and some basic laws of physics (which we cover in the book), we can understand how we know that the Big Bang occurred 14 billion years ago, and that a thimbleful of white dwarf material has as much mass as five elephants. Our goal was not simply that our students be told about the wonders of the universe, but that they learn the tools to understand *how* we know what we know. We aimed to empower our students to apply quantitative and physical reasoning to their everyday lives, to understand the basis behind the very real issues that face us as citizens of the world.

In mathematics, physics, and indeed astronomy, equations give us a way to realize, and understand, the relationship between concepts that we may not have previously known were connected. Newton's law of gravity connects the force on a dropped pencil to the mass and radius of the Earth. Planck's radiation law gives us a relationship between the temperature and size of a star and its luminosity. And Einstein's famous formula, $E = mc^2$, tells us that energy E and mass m are intimately related, and that in a fundamental sense, they are two aspects of a common phenomenon, called "mass-energy."

Welcome to the Universe was not written as a textbook per se, but rather was meant to be read as you read any book for pleasure, albeit perhaps with a pad of paper by your side if you want to follow along with some of the calculations. However, the book *can* be used in a classroom setting, perhaps in conjunction with a more traditional textbook. It is also ideal for use in a "flipped" classroom, where the lectures are on videos and the teacher works with groups of students on solving problems in class. We thus make this volume of problems available, which ask you, the reader and student, to grapple directly with the quantitative and conceptual aspects of the subject. Almost all the problems here were used in this course and other courses we have taught at the same level. These problems range from the fairly easy to the challenging, but they all require no mathematics beyond high-school algebra. The problems are not just exercises in applying astronomical formulas but are designed to give insight into specific astrophysical problems. We hope that in the process of solving these problems, you will gain both some facility in the quantitative techniques of our field and also experience that "ah-ha" moment, when you measure the expansion rate of the universe from direct observations, or repeat the calculation astronomers did when neutrinos were first discovered that had been emitted from a supernova 150,000 light-years away.

We have put the solutions to these problems in the back of the book. We encourage you to attempt these problems first before looking at the solutions. But you will see that the solutions go into quite a bit of pedagogical detail, and they often give additional insights and background material that goes beyond the initial problem. Even if you are able to solve a problem fully on your own, you are likely to gain new insights by reading the solutions. Of course, an easy and fun way to read this book is to read each problem and its solution together, to see the answers and the techniques to solve each problem revealed.

Many of these problems can be used as is in a classroom setting, or as inspiration for similar problems that the teacher or professor can devise. We are rather fond of multi-part problems that build on one another, but these can be reduced considerably as the instructor finds appropriate. The underlying concepts do sometimes repeat in these exercises. For example, we have many problems exploring the range of densities in the universe, from the black hole limits set by our lack of understanding of quantum gravity (yes, you can calculate such a thing using just high-school algebra) to the incredibly low density of the universe as a whole. We encouraged our students to explain their answers in full (i.e., in words) in their homework, to demonstrate that they understood the context and content of what they were doing and were not simply plugging into a formula. Each instructor will need to find the appropriate balance of English and mathematics for their students. In that spirit, there are some more qualitative questions, as well as essay questions throughout the book.

This problem book starts with some general musing and advice on problem-solving skills and tools. We assume that you are familiar with scientific notation, the way we write the really large and really small numbers that come up so often in our subject. The problems are ordered according to the chapter structure of *Welcome to the Universe*, giving you guidance on which problems to attempt as you read the book. In a separate section following the problems, we tabulate some of the most useful numbers and equations used throughout these problems (and a few that are not); you should take these as given in these problems, unless we direct you otherwise. One of the themes throughout is that algebra is often easier than arithmetic, and in these problems we put a lot of emphasis on algebraic techniques to simplify problems before plugging in numbers. Many of these problems require answers to "astronomical accuracy" (i.e., to a rough precision only), and we demonstrate a style of calculator-free arithmetic that can quickly and accurately get you to the answer.

We thank all those who have contributed to these problems. Our colleagues Chris Chyba, Joe Patterson, Anatoly Spitkovsky, Jenny Greene, and David Spergel have used many of these problems in teaching the Princeton introductory astronomy course and have helped us refine them. Chris Chyba contributed a number of the problems here; we mark them as such. Jeremy Goodman, David Spergel, Vera Gluscevic, and Mariangela Lisanti gave helpful comments and insights on several problems. We worked with a significant number of teaching assistants, both graduate and undergraduate, who helped refine the solutions and did the lion's share of grading homework assignments and exams. We thank our wonderful editor Ingrid Gnerlich for her continuing faith in this book and for all her support. But most of all, we thank the thousands of students who have taken our course at Princeton, who continue to inspire us with their energy, their curiosity, and their insightful questions.

Michael A. Strauss
March 2017

MATH TIPS

Significant Figures

In mathematics courses that you have taken in high school and in college, you were taught to think of numbers as absolute and precise quantities. Thus, for example, if you were asked to divide 10 by 3, the correct answer is $3.3333\ldots$, or $3.\bar{3}$, where the $^-$ sign over the 3 means that the 3s continue on forever.

However, in science (and especially astronomy), numbers are often not known precisely. We refer to the number of significant figures a number has: this means the number of digits it is written with in scientific notation. For example, the number 5.2987×10^{-11} has 5 significant figures, and the number $4 \, (= 4 \times 10^0)$ has a single significant figure. In writing a number with a certain amount of significant figures, we are making a statement about the precision with which we know the number. That is, when we write a number like 5.2987×10^{-11}, we are saying that we have confidence it isn't as large as 5.2988×10^{-11}, or as small as 5.2986×10^{-11}. This degree of certainty or uncertainty carries through in the calculations we do with this number.

For example, suppose we are told that the nearest star is about 4 light-years away, and are asked to convert this number to kilometers. We know that 1 light-year is about 9.46×10^{12} km, so the calculation seems straightforward:

$$4 \text{ light-years} = 4 \text{ light-years} \times \frac{9.46 \times 10^{12}\,\text{km}}{1\,\text{light-year}} = 3.784 \times 10^{13}\,\text{km}.$$

This is what your calculator would display, but this is not correct. The problem is that the number, 4 light-years, is given to you with only a single significant figure. That is, from the statement of the problem, we only know that the distance to the nearest star is between 3.5 and 4.5 light-years (i.e., between 3.3×10^{13} km and 4.3×10^{13} km). So it is misleading and wrong to give an answer with four significant figures; it implies that you know the answer much more precisely than you actually do.

The right way, then, to handle any calculation involving multiplication and division is to limit the precision of your answer to that of the input number with the least number of significant figures. In this case, 4 light-years has exactly one significant figure, and thus the answer should have a single significant figure. The correct answer in this case is 4×10^{13} km.

Note that if we had told you that the distance to the nearest star was 4.00 light-years, we would be giving you three significant figures, and you should give the resulting distance in kilometers to three significant figures.

In doing astronomy, we will often find that our numbers are known rather imprecisely, to only one or two significant figures. This makes calculating quite a bit easier than it would be otherwise. When you are doing arithmetic to one significant figure (as you will do in a lot of these problems), things really become easier, and you are allowed to make approximations that would horrify your high-school math teacher, such as $4/3 \approx 1, 3 \times 3 \approx 10$, and so on. Again, take the example of 4 light-years. We know already that the final answer will have only one significant figure, so we are justified in rounding the number of kilometers per light-year to a single significant figure, 1 light-year $= 1 \times 10^{13}$ km. Now the calculation becomes so easy that we can do it without a calculator:

$$4 \text{ light-years} = 4 \text{ light-years} \times \frac{1 \times 10^{13}\,\text{km}}{1\,\text{light-year}} = 4 \times 10^{13}\,\text{km}.$$

This is easier than the first calculation, less prone to error, and moreover gives us an answer with the right number of significant figures. In the solutions, we will give many examples of doing such arithmetic without a calculator. However, for longer calculation, it sometimes is a good idea to keep one extra significant figure during the calculation and round only at the end. To take a simple example, you might be tempted, in calculating 2.4×4, to round the first number down to 2, giving the result of 8. Doing the calculation exactly gives 9.6, and then rounding gives 10.

Let's think a bit more deeply about what the point of significant figures are: they really express the uncertainty we have for a given number. Thus suppose we're asked to add 8 and 6. The single significant figure in the two cases implies uncertainties (roughly) of about one in each case. The sum of 8 and 6 is 14; should we round the result down to a single significant figure of 10? In this case, no, because we know the uncertainty in the sum is roughly the sum of the uncertainties of the two individual numbers (i.e., about two).[1] Rounding from 14 to 10 is a change substantially larger than the uncertainty in the sum, so in this case, it is best to keep two significant figures.

We won't always write numbers in scientific notation, but usually by context you can understand the number of significant figures. For example, if we tell you that a star has a temperature of 10,000 degrees, you can assume that this number is known to one or perhaps two significant figures, unless we explicitly tell you otherwise.

Thus as rules of thumb:

- When doing calculations with numbers with one or two significant figures, you can do your arithmetic without a calculator as illustrated above and quote your answer to one (or occasionally two) significant figures. Most of the calculations you will do in this book will be of this type.
- When you have to keep track of more significant figures, it may be easiest to do the calculation "exactly" on a calculator, and then round to the appropriate number of significant figures in the end.

Often in astronomy, we need to get a rough, "order-of-magnitude" estimate of a quantity, and many of the problems are in this spirit. For example, in one problem we will ask whether 1 ton of white dwarf material will fit into a matchbox; for a problem like this, one significant figure will give you a clear answer to this question!

The above rules about significant figures are a bit trickier when dealing with subtraction. We will occasionally see problems in which two large numbers need to be subtracted from one another:

$$4,000.001 - 4,000.000 = 0.001.$$

1 There is a mathematically correct way of combining uncertainties that gives a somewhat smaller quantity than this, but that is beyond the scope of this book.

In this example, if we rounded the numbers off before doing the subtraction, we would find ourselves with an answer of zero, which may miss the point of the problem. In another example, in the relativity section of *Welcome to the Universe*, we will be dealing with objects moving *very* close to the speed of light, c, and we write their speed as, say, $v = 0.999999999999c$. You will of course be tempted to round this number to c, but as we'll see, the physically significant quantity (and in an interesting sense, the one we'll actually measure) is the difference between the c and the actual speed v (i.e., $10^{-12}c$ in this example), which we actually know to a single significant figure. We will give appropriate hints in the problems in which these issues come up.

Algebra and Arithmetic

Even though you learned arithmetic before algebra in school, it is often true that arithmetic is more difficult than algebra. A general rule that will hold you in good stead when doing these problems is to simplify calculations as much as possible using the techniques of algebra *before* doing any arithmetic. Here is an example that will come up in the problems. You are asked to take the ratio of the fourth root of the luminosities of two stars, whose values are 3.2×10^{27} and 2×10^{26} Joules/sec, respectively:

$$\frac{(3.2 \times 10^{27}\,\text{Joules/sec})^{1/4}}{(2 \times 10^{26}\,\text{Joules/sec})^{1/4}}.$$

At this point, you may be tempted to pull out your calculator and find that

$$(3.2 \times 10^{27}\,\text{Joules/sec})^{1/4} = 7.52 \times 10^{6}\,(\text{Joules/sec})^{1/4};$$

$$(2 \times 10^{26}\,\text{Joules/sec})^{1/4} = 3.76 \times 10^{6}\,(\text{Joules/sec})^{1/4},$$

then take their ratio, worrying terribly all along whether you're dealing with significant figures correctly, and what these strange units of $(\text{Joules/sec})^{1/4}$ mean... But life is much simpler when you realize that the ratio of powers is the power of the ratio; that is, you can write the calculation as

$$\left(\frac{3.2 \times 10^{27}\,\text{Joules/sec}}{2 \times 10^{26}\,\text{Joules/sec}}\right)^{1/4} = 16^{1/4} = 2.$$

Look Ma, no calculator! And the units canceled as well. We will often set up problems to be done easily with tricks like this.

The problems will often use ratios to do calculations with quantities that are proportional to one another. Suppose we tell you about two stars that have equal surface temperatures, but star A has twice the radius of star B. We ask you to calculate the ratio of their luminosities. In *Welcome to the Universe*, you will learn that the luminosity L of a star of a given temperature is proportional to the square of its radius R:

$$L \propto R^2,$$

where the $\propto$ symbol means "proportional to." What this really means is that there is a constant C, such that

$$L = CR^2.$$

Let us use this to determine the ratio of the two quantities:

$$\frac{L_A}{L_B} = \frac{CR_A^2}{CR_B^2} = \frac{R_A^2}{R_B^2} = \left(\frac{R_A}{R_B}\right)^2.$$

Note that the constant C cancels out here; we don't need to know it. But writing down the constant as C allowed us to remember what the proportionality meant. Also note the trick, like that described above, of turning a ratio of squares into the square of a ratio. Now we are already told that $R_A/R_B = 2$, so the calculation is easy: the ratio of the luminosities is the square of this, or 4.

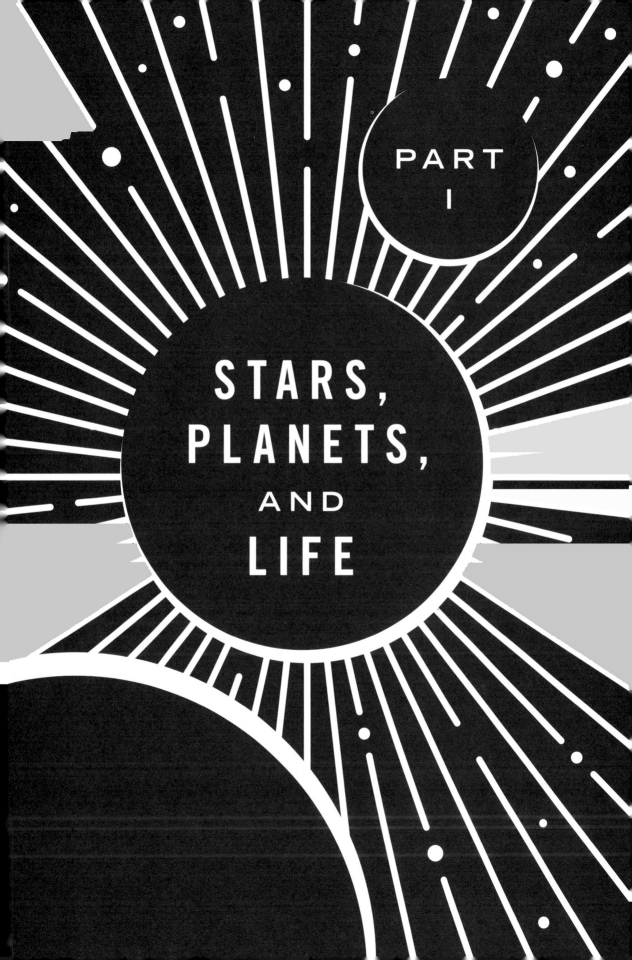

PART
I

STARS,
PLANETS,
AND
LIFE

in AU and in light-years. Hint: *Remember that the parallax is traditionally defined as* **half** *the angle that the position of a star changes as Earth goes from one side of its orbit around the Sun to the other.*

6. Looking out in space and back in time

The speed of light is 3.0×10^5 km/sec. The distance between Earth and the Sun (1 Astronomical Unit, or AU) is 150 million kilometers. Give all answers to the correct number of significant figures.

6.a You look at your friend, 30 feet (10 meters) away. How far back in time are you seeing her? Express your answer in nanoseconds. The light entering your eyes from your friend is ambient light in the surroundings that reflects off her.

6.b Electrical engineers are always looking for ways to increase the speed of computer chips. One way to allow computers to perform more operations per second is simply to reduce the distance between components, so that the electrical circuits are shorter. The electrons that "flow" from one transistor to the next in your computer travel at close to the speed of light. In modern chips, these distances are impressively small, in 2017 approaching 10 nanometers (one hundredth of a micron). How long, in seconds, does it take the electrons to "flow" from one transistor to the other? Please use scientific notation.

6.c How far back in time are we looking when we observe the Orion Nebula, which is 1,500 light-years away? Express your answer in years.

7. Looking at Neptune

All electromagnetic radiation (including radio waves) travels at the speed of light. In 1989, the Voyager II spacecraft flew by the planet Neptune. Neptune is on a roughly circular orbit around the Sun of radius 30 AU, while Earth is on a circular orbit of radius 1.00 AU (of course!). (For this problem, at the level of precision asked for, the approximation that these planetary orbits are circles lying in the same plane is a good one.)

7.a How far back in time are we seeing Neptune, when Neptune is farthest from Earth in the two planet's orbits? Give your answer in hours and minutes.

7.b How far back in time are we seeing Neptune, when it is *nearest* to Earth in the two planet's orbits? Give your answer in hours and minutes.

8. Far, far away; long, long ago

Problem suggested by Chris Chyba.

Because the speed of light is finite, looking at anything means seeing the thing as it was some time in the past. This usually doesn't matter in daily life on Earth, but it is often important in astronomy and spacecraft engineering, or for communication among galactic civilizations. We'll explore this question here.

8.a Earth started transmitting powerful radio signals in the 1930s, and these signals travel out to space at the speed of light. Consider a radio broadcast sent in 1936. If there were extraterrestrial civilizations listening for radio signals drifting out into the Galaxy, and they were to pick up this broadcast, out to what distance from our solar system could these civilizations have already detected our presence? Give your answer in light-years; round to whole numbers.

8.b The Arecibo radio telescope in Puerto Rico (one of the largest radio telescopes on Earth) can transmit radio signals (as a radar transmitter) as well as receive them. Arecibo is so sensitive that, if there were a duplicate Arecibo on another planet (call it Zyborg) half-way to the center of the Galaxy, and Zyborg's Arecibo were pointed at Earth, it could pick up a transmission from Arecibo, Puerto Rico. The Sun is about 25,000 light-years from the center of the Milky Way. How long would it take a signal sent from Arecibo, Puerto Rico to reach Arecibo, Zyborg? Again, express your answer in years, to the appropriate number of significant figures.

8.c Suppose you were trying to have a conversation with an astronaut on the Moon. The Moon is 384,400 kilometers from Earth. If you sent a radio signal to an astronaut on the Moon, and she replied immediately, how long would be the gap between when you sent the signal and you received your answer? Give an answer to two significant figures.

8.d Now imagine that you are an engineer for the Mars Exploration Rover scooting around on the surface of Mars. Suppose you need to send it an emergency transmission to prevent it from driving into a ditch. Mars is on a roughly circular orbit of radius 1.5 AU around the Sun. Earth is also on a circular orbit in the same plane, of radius 1.0 AU, of course. (For these two planets the approximation of their orbits as circles is a good one at the level of precision asked for in this problem.) About how long would it take your message to reach the rover when Mars is in

2

FROM THE DAY AND NIGHT SKY TO PLANETARY ORBITS

12. Movements of the Sun, Moon, and stars

Explain where on Earth, and when (time of day and/or time of year, as relevant), each of the following situations occurs. It is possible in some of these cases that the situation described never happens. In each case, give a full explanation, using diagrams where helpful.

12.a The Sun is above the horizon for 24 hours straight.

12.b A half-illuminated Moon is in the sky (i.e., above the horizon) at the same time as is the Sun (i.e., during the day).

12.c The planet Mars is seen transiting (i.e., passing in front of) the Sun.

12.d None of the visible stars ever appears to set (i.e., drop below the horizon).

12.e Polaris is directly overhead.

13. Looking at the Moon

13.a On a certain night last year, you went outside and saw a full Moon on the eastern horizon. Approximately what time was it?

13.b Next to that full Moon was one of the bright planets. It was either Venus or Jupiter. Which of the two must it be?

13.c Would residents of the far side of the Moon (i.e., the side of the Moon currently facing away from Earth) ever see Earth? How long would the "day" (i.e., the time from one sunrise to another) be for them (or would they ever see the Sun at all)?

14. Rising and setting

Because of Earth's rotation, the stars, the Sun, and the Moon all rise in the east and set in the west, rotating around an axis that points to the North Pole. Answer the following questions with full explanations.

14.a Does an observer at the North Pole also see stars rise and set?

14.b Does an observer at the North Pole see the Sun rise and set in a single day? Over the year? One often sees travel brochures for Alaska billing it as "The Land of the Midnight Sun." What does this mean in this context? *Hint: Is the Sun fixed in the sky relative to the stars?*

14.c The bright star Polaris, or the North Star, lies almost exactly at the Celestial North Pole. That is, if one imagines drawing a line that extends from the South to the North Pole of the Earth and out into space, it points directly to this star. Where does this star appear to an observer at the North Pole? Does your answer depend on the time of night or time of year?

14.d Where does Polaris appear to an observer living in Quito, Ecuador (i.e., on Earth's equator)? Does your answer depend on the time of night or time of year?

14.e Using a telescope, it is possible to see at least bright stars during the day. Are there any stars that are visible from Princeton (roughly 40° north latitude) 24 hours a day (at least if it remains clear)? That is, are there stars which never set as seen from Princeton? If the answer is yes, explain where these stars are in the sky. Please be as quantitative as possible. A diagram may be helpful.

15. Objects in the sky

There is a lot that you can learn by looking at the night sky.

15.a You stumble outside after a marathon session of working on your astronomy homework, wondering what time it is. You look up in the sky, and see a beautiful full Moon high in the sky. Roughly what time is it? Explain your answer.

15.b An object in the sky is termed "circumpolar" if it stays above the horizon for 24 hours (i.e., it never rises or sets). Is the Moon ever circumpolar as seen from the equator? Explain your answer.

15.c Is the Moon ever circumpolar as seen from the North Pole? Explain your answer.

15.d From a mid-latitude Northern Hemisphere site, you see a lunar eclipse that takes place at midnight. Was the Moon

- Rising on the eastern horizon,
- Setting on the western horizon,
- High in the sky, or
- Not visible because it was below the horizon?

Explain your answer. A diagram may be useful.

16. Aristarchus and the Moon

The ancient Greek philosopher Aristarchus realized that he could use the angle between the Moon and the Sun when the Moon was half-illuminated to determine the ratio of the Earth–Moon distance to the Earth–Sun distance. In this problem, we will follow his logic quantitatively.

In the accompanying figure 1, E, M, and S refer to Earth, the Moon, and the Sun, respectively. The Moon is indicated at first quarter (i.e., when it appears exactly half-full as seen from Earth), where Earth, the Moon, and the Sun make a right angle. Similarly, the third quarter is at point M'. Because the Sun isn't infinitely far away, the points of first and third quarter are not exactly opposite each other on the Moon's orbit. In particular, there is a nonzero angle between the Earth–Moon line at first quarter, and the line perpendicular to the Earth–Sun line (i.e., the line from E to A).

16.a Use your knowledge of high-school geometry to demonstrate that the angle ESM is equal to the angle AEM.

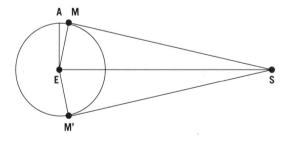

Figure 1. Figure for problem 16. This is not drawn to scale. E represents Earth, M and M' the Moon, and S the Sun. The circle represents the Moon's orbit. The Moon is indicated when it is at first and last quarter, and is half-illuminated as seen from Earth. At these points, the angles EMS and EM'S are right angles.

16.b Use the small-angle formula to estimate this angle in terms of the distance $\overline{EM}$ from Earth to the Moon, and the distance $\overline{ES}$ from Earth to the Sun. Express the angle both in degrees and radians.

16.c You should have found that the deviation between the angle SEM and $90°$ is very small. Aristarchus determined (incorrectly) that that angle was $3°$. Given that the angular diameters of the Moon and the Sun are the same, what did he conclude about the ratio of the physical diameter of the Sun to that of the Moon?

17. The distance to Mars

You and a friend set out to measure the distance to Mars via parallax. After synchronizing your watches, you position yourself at two points separated by 1,000 kilometers on Earth. The line separating the two of you is perpendicular to the line between Mars and the center of Earth. At exactly the same moment in time, you each measure the angular position of Mars relative to the background stars. You find that the two positions differ from each other by 2 arcseconds. How distant is Mars from Earth at that moment? Express your answer in kilometers. *Hint: Here, "parallax" refers to the general phenomenon whereby the angle at which we see a distant body depends on our position. This is in contrast to the specific usage of the term to refer to the apparent change of position of stars due to the Earth's orbit around the Sun.*

18. The distance to the Moon

You and a friend set out to measure the distance to the Moon via parallax. Here we're not talking about the parallax due to the motion of Earth around the Sun but from observing from two different spots on Earth. After synchronizing your watches, you position yourself at two points on Earth separated by 200 kilometers. The line separating the two of you is perpendicular to the line between the Moon and the center of Earth. You each measure the angular position of a specific point on the edge of the Moon relative to the background stars at exactly the same time. By how much do you expect the positions the two of you measure to differ? Express your answer in arcseconds.

19. Masses and densities in the solar system

In astronomy, the densities of objects cover a truly enormous range. In this problem, you are going to explore part of this range.

19.a The mass of the Sun is roughly 3×10^5 times the mass of Earth, and the mass of Earth is roughly 6×10^{24} kilograms. The Sun is observed to subtend an angle on the sky of half a degree. The distance from Earth to the Sun is 1 Astronomical Unit (abbreviated AU), which is roughly 1.5×10^8 kilometers. Given this information, calculate the radius of the Sun and its density relative to water (which has a density of 1,000 kilograms per cubic meter). *Hint: All the numbers you need for this calculation are given here. Try to do the calculation without a calculator; only an approximate answer is needed.*

19.b Neptune, the outermost planet, has an orbit with a radius of roughly 30 AU. What is the mean density of the solar system out to this radius? That is, if you were to take all the mass of the solar system and spread it out evenly within a sphere of radius 30 AU, what would its density be? Assume that the mass of the planets is negligible relative to the Sun (see part **d**).

19.c The most distant comets are thought to come from a spherical cloud (the "Oort cloud," named after Jan Oort (1900–1992), who first hypothesized its existence) centered on the Sun with a radius of roughly 1 light-year. We could argue that this represents the true outer boundary of our solar system. Calculate the mean density of the solar system within a sphere of this radius. The total mass of the comets in the Oort cloud is estimated to be a few times the mass of Earth (i.e., tiny compared to the Sun).

19.d Is the approximation that the mass of the planets is negligible a valid one? Use the masses of the planets (tabulated in table 9.1 of *Welcome to the Universe*) to calculate the ratio of the total mass of the planets to the mass of the Sun, and answer this question.

3

NEWTON'S LAWS

20. Forces on a book

20.a A book is sitting on a table on Earth. Is it at rest? It is moving at constant velocity? Identify all the relevant forces acting on the book, and apply Newton's second law of motion to answer the question of whether the book is being accelerated.

20.b Did you include the gravitational effect of the Sun in your calculation? Should you have? Now answer the question again: is the book moving at constant velocity?

20.c Everything in the universe which has mass gravitates. Which is larger, the gravitational force imparted on the book by the Sun, or that from all the other planets, and Earth's Moon, combined? For definiteness, consider a situation in which the Sun, all the planets, and the Moon, appear in the same direction in the sky, as seen from Earth.

21. Going ballistic

A satellite in low Earth orbit, 400 kilometers above Earth's surface (such as the Hubble Space Telescope), takes approximately 90 minutes to complete one orbit.

21.a How fast is the satellite traveling? Express your result in meters per second and kilometers per hour.

21.b Use the information you've been given, together with Newton's law of gravity, to calculate the mass of Earth.

22. Escaping Earth's gravity?

At what distance along the line separating Earth and the Moon are the gravitational pulls of the two equal and opposite? Could one say that beyond that point, one is "free of Earth's gravity?"

23. Geosynchronous orbits

A satellite in geosynchronous orbit (GEO) orbits Earth once every day. A satellite in geostationary orbit (GSO) is a satellite in a circular GEO orbit in Earth's equatorial plane. Therefore, from the point of view of an observer on Earth's surface, a satellite in GSO seems always to hover at the same point in the sky. For example, the satellites used for satellite TV are in GSO, so that satellite dishes on Earth can be stationary and need not track their motion through the sky. Take a look; in the Northern Hemisphere, you'll notice all satellite dishes on people's houses point toward the equator (i.e., South).

How far above Earth's equator (i.e., above Earth's surface) is a satellite in GSO? Express your answer in kilometers and in Earth radii.

24. Centripetal acceleration and kinetic energy in Earth orbit

Problem suggested by Chris Chyba.

In January 2007, China conducted an antisatellite weapon (ASAT) test in low-Earth orbit, in which it collided an interceptor with one of its own aging weather satellites, the Fengyun 1-C, orbiting at an altitude of 850 km above Earth's surface. The previous ASAT test had been conducted by the United States in 1985 at an altitude of 350 km. The Soviets also conducted ASAT tests from the 1960s to the 1980s, though their system did not rely on collision but rather got the two satellites close to each other and then exploded the interceptor.

24.a Derive an algebraic expression for the speed v of an object in circular orbit of radius r around Earth, in terms of the mass of Earth $M_\oplus$, r, G, and numerical constants. You can do this either by starting with Newton's form of Kepler's third law or by using Newton's law of gravitation and his second law of motion.

24.b Assume that the Fengyun 1-C was in a circular orbit prior to its destruction. What was its speed? Express your answer in kilometers per second. What was its orbital period? Express your answer in minutes. *Hint: Think hard about what you want to use for the radius of the orbit!*

24.c Successful ASAT tests result in the production of a great deal of orbital debris, including "large" debris, defined as pieces with size greater than 10 cm. China's 2007 test produced over 1,700 pieces of large debris that are now being tracked by NASA. (The 1985 U.S. test produced a comparable amount of large debris. The orbits of debris in low-Earth orbit slowly decay due to the drag of the tenuous atmosphere at that altitude, and because the U.S. test was at much lower altitude where the atmosphere is thicker, the orbits decayed much more quickly. The last large piece of debris from the 1985 test fell to Earth in 2004.) Assume that a typical piece of large debris is a square of dimensions 10 cm by 10 cm with thickness 2 mm and is made of aluminum with a density 2.7 grams per cubic centimeter. What is the mass in kilograms of a typical piece of large debris?

24.d Assume that after the destruction of the target satellite, a typical piece of debris is still moving with the orbital velocity found in part **b**. (Some pieces will be moving faster and some slower, but this is not a bad approximation for an average velocity after the intercept; most of the debris remains in orbits broadly similar to that of the original satellite.) What is the kinetic energy, in units of Joules, of a typical piece of large debris?

24.e The energy per kilogram released in an explosion of the high explosive TNT is 4.2×10^6 Joules/kg. Suppose that a piece of large debris were to hit another satellite with the velocity you used in part **d**. Which would be more destructive: a collision with the piece of debris you considered in part **d**, or an explosion from a piece of TNT of the same mass? To answer this question, calculate the ratio of the energy from a collision and from the explosion.

25. Centripetal acceleration of the Moon and the law of universal gravitation

This problem examines the way in which Newton claimed, late in his life, to have first thought of the law of universal gravitation.

25.a The Moon orbits around Earth in nearly a circle of radius 384,000 kilometers; we'll ignore the small ellipticity of the orbit in this problem. How many Earth radii is this distance? It takes 27.3 solar days (1 month) to complete this orbit. What is the velocity (in m/sec) of the Moon in its orbit? What is the centripetal acceleration (in m/sec^2) of the Moon in its orbit?

25.b By Newton's second law of motion, the force required to give a body of mass m an acceleration a is ma. For a body moving in a circle of radius r at constant speed v, that acceleration is v^2/r. Given this, what must the relationship between the force and radius be to also satisfy Kepler's third law, which states that r^3 is proportional to the square of the period P? We are looking for a proportionality between the force and a simple function of radius.

25.c If the centripetal acceleration of the Moon is really caused by Earth's gravity, then this acceleration should fall off in proportion to the square of the distance from Earth. Use the distance of the Moon in Earth radii to predict what the gravitational acceleration at Earth's surface (i.e., 1 Earth radius from the center) should be. Compare the result with the measured acceleration of 10 m/sec^2 for an apple falling from a tree.

26. Kepler at Jupiter

In this problem, we will check Kepler's third law explicitly. We'll examine the orbits of the four moons that Galileo discovered orbiting Jupiter: Io, Europa, Ganymede, and Callisto. The semi-major axes a and periods P of their orbits are as follows:

Moon	Semi-major axis (km)	Period (days)
Io	4.216×10^5	1.769
Europa	6.709×10^5	3.551
Ganymede	1.070×10^6	7.155
Callisto	1.883×10^6	16.69

In this problem, you will confirm that they scale the way Kepler predicted: $P^2 = Ca^3$ (where C is a constant of proportionality). A calculator would be appropriate for this problem.

26.a First, calculate C using Europa, first using MKS units (time measured in seconds, and distances measured in meters) and then using units in which time is measured in years and distance is measured in Astronomical Units.

26.b Using the value of C you calculated using MKS units, confirm that Kepler's third law holds for Io, Ganymede, and Callisto.

26.c For the orbit of Earth around the Sun, what is the value of the proportionality constant C, in units of AU and years?

26.d Using units of AU and years, what is the ratio of the constant of proportionality C you found in part **a** for the Jupiter system to the value of C that holds for Earth's orbit around the Sun? Also calculate the ratio of Sun's mass to Jupiter's mass; how do these two ratios relate to one another? Why?

27. Neptune and Pluto

Problem suggested by Chris Chyba.

The planet Neptune, the most distant known major planet in our solar system, orbits the Sun with a semi-major axis $a = 30.066$ AU and an eccentricity $e = 0.01$. Pluto, the next large world out from the Sun (though much smaller than Neptune) orbits with $a = 39.48$ AU and $e = 0.250$.

27.a To the correct number of significant figures, given the precision of the data in this problem, how many years does it take Neptune to orbit the Sun?

27.b How many years does it take Pluto to orbit the Sun?

27.c Take the ratio of the two orbital periods you calculated in parts **a** and **b**. You'll see that it is very close to the ratio of two small integers; which integers are these? Thus the two planets follow the same pattern with respect to each other repeatedly, and thus their mutual gravitational attraction is reinforced over time, which keeps them locked in this pair of orbits. This is an example of an orbital resonance. Other examples in the solar system can be found among the moons of Jupiter, and between the moons and various features of the rings of Saturn.

27.d What is the aphelion distance of Neptune's orbit? Express your answer in AU.

27.e What are the perihelion and aphelion distances of Pluto's orbit? Is Pluto always farther from the Sun than Neptune is?

28. Is there an asteroid with our name on it?

Problem suggested by Chris Chyba.

In the Hollywood movie *Armageddon*, Bruce Willis saves the world at the last minute by blowing up an asteroid headed directly toward Earth. In real life, there are indeed asteroids with the potential to impact Earth; in this problem, we'll explore their properties and figure out a more realistic way to avoid Doomsday.

28.a An asteroid in orbit around the Sun is defined here to be an "Earth-crosser" if its perihelion (its closest approach to the Sun) is within 1 AU, and its aphelion is outside 1 AU, and therefore it has the potential to impact Earth. The asteroid 2002 CY9 has a period of 2.12 years and an orbital eccentricity of 0.50. Calculate the semi-major axis of the orbit, as well as the perihelion and aphelion distances, in AU. Is it an Earth-crosser?

28.b A typical Earth-crossing asteroid like 2002 CY9 is traveling at a speed of roughly 30 km/sec relative to Earth, so if it were to hit Earth, it would cause an enormous amount of damage. There are several sky surveys under way to find such potentially hazardous asteroids; mapping their orbits allows us to determine whether they are likely to impact Earth, giving us the opportunity to try to prevent the collision. Imagine that you have just discovered an asteroid that will hit Earth 20 years from now. You are given the task to design a spacecraft that will impact the asteroid and change its orbit just enough to avoid hitting Earth. So you send a massive rocket on a collision course with the asteroid, hitting it so hard that you change its velocity by 2 cm/sec. By how much does the position of the asteroid change in the 20 years before it comes close to Earth? You can do this calculation very crudely by simply asking how much distance a 2 cm/sec nudge corresponds to over 20 years. Is this enough to cause it to miss Earth?

29. Halley's comet and the limits of Kepler's third law

Consider Halley's comet, which orbits the Sun in a very eccentric orbit, with an aphelion (i.e., the point on the orbit farthest from the Sun) out beyond the orbit of Neptune. Every 76.1 years it swings in toward the Sun, crossing the orbits of most of the planets. As it gets close to the Sun, its ice begins to sublime (i.e., it goes directly from solid to gas phase), shedding gas and dust and leading to jets of debris blasting off of its nucleus. The semi-major axis of Halley's orbit is 17.8 AU.

29.a Show that for Halley's comet, Kepler's third law is not exactly obeyed! A calculator will be useful here.

29.b Suggest a reasonable hypothesis for why Halley's comet does not obey Kepler's third law as precisely as do the objects in problem 26. *Hint: Is the Sun's gravity the only force acting on the comet?*

30. You cannot touch without being touched

Let's put this poetical phrasing of Newton's third law of motion to work in understanding the effect of planetary motions on the Sun. Jupiter is the most massive of all the planets. It has a roughly circular orbit around the Sun of radius 5.2 AU.

30.a What is the period of Jupiter's orbit in years?

30.b What is the speed at which Jupiter moves? Give your answer in kilometers per second.

30.c The mass of Jupiter is just about 1/1,000 that of the Sun. Ignoring the effects of the other planets (Jupiter is, after all, by far the most massive of the planets), Newton's third law of motion says that the center of mass of the Sun-Jupiter system will stay fixed as Jupiter moves; therefore, the Sun must move in response to Jupiter's motion. Calculate how far the Sun moves in response to the motion of Jupiter over half its orbit. Then calculate how fast it moves.

31. Aristotle and Copernicus

Describe the Aristotelian worldview, emphasizing its description of the physical nature of, and laws that govern the motions of, heavenly bodies relative to those on Earth. Explain how this differs from our current scientific understanding of the universe, and explain the importance of the discoveries of Galileo Galilei, Nicolas Copernicus, Johannes Kepler, Isaac Newton, and Cecilia Payne-Gaposchkin in overturning the Aristotelian worldview.

4–6

HOW STARS RADIATE ENERGY

32. Distant supernovae

In 1987, a supernova exploded in the Large Magellanic Cloud, a small galaxy near the Milky Way at a distance of 150,000 light-years. At its peak brightness, it was easily visible to the naked eye, about 1/10 the apparent brightness of the star α Centauri (α Cen, one of the nearest stars to us, at 4 light-years away). By what factor was the luminosity of the supernova at its peak larger than that of α Cen?

33. Spacecraft solar power

Problem suggested by Chris Chyba.

NASA's Mars Reconnaissance Orbiter (MRO) carries six instruments to study the Martian atmosphere and surface, including a ground-penetrating radar to search for water beneath the planet's surface. You can find more information (not necessary for this problem) at http://www.nasa.gov/mro.

MRO communicates with Earth using a 10-foot-diameter dish antenna and a transmitter powered by 100 square feet of solar panels. Modern solar panels have conversion efficiencies of 20% (i.e., the panel converts 20% of the sunlight incident on it into electrical power—the other 80% is lost).

33.a Given the surface temperature of the Sun (6,000 K) and its radius (7.0×10^5 km), calculate its blackbody luminosity, in watts (Joules/sec).

33.b Mars is 1.5 AU from the Sun. Calculate the brightness of the Sun at Mars' distance (i.e., the solar flux on Mars' surface) expressed in watts per square meter.

33.c Using the approximation (good to about 10%) that 1 meter = 3 feet, calculate how much electrical power will be available to the MRO transmitter, assuming that the MRO solar panels are facing the Sun and that they have a conversion efficiency of 20%. Express your result in watts.

33.d Suppose now that a spacecraft identical to MRO were launched (presumably using a bigger rocket!) to travel to Saturn's moon Titan. How much power (in watts) would the MRO at Saturn have available for its transmitter? Saturn orbits at 9.6 AU from the Sun. For comparison, a typical incandescent light bulb in your home uses about 100 watts of electrical power. What would be the ratio of MRO transmitter power at Saturn to that of a 100-watt light bulb?

34. You glow!

Calculate the luminosity of your body due to blackbody radiation. Be explicit about the assumptions and approximations you make. *Hint: We are looking for an approximate answer.* Give the answer in watts (Joules/sec), and compare with a typical lightbulb (100 watts). Given all this energy people are radiating, why is it that we can't see each other in a room with the lights off?

35. Tiny angles

In astronomy, we often refer to angles on the sky. The relationship between the angular distance θ between two objects on the sky, their physical separation s from each other, and their distance d from us is given by $\theta = s/d$, if the angle θ is measured in radians.

35.a A star at a distance of 300 light-years from Earth is moving perpendicular to our line of sight at 300 kilometers per second relative to us. How long will we have to wait until it appears to have moved 1 arcsecond relative to more distant stars? Express your answer in years. *Hint: 300 kilometers per second is much larger than the speed of our orbit around the Sun; thus you can neglect the effect of the latter.*

35.b We found that this star moves an arcsecond on the sky in a bit more than a year. This is quite a fast-moving star. The world record for apparent motion on the sky is held by a star known as Barnard's Star, which moves at 10 arcseconds per year. Barnard's star is 6 light-years away. What is its speed in the plane of the sky? Express your answer in kilometers per second.

36. Thinking about parallax

Here we are going to think about how the apparent positions of stars in the sky are affected by the orbit of Earth: the parallax effect.

36.a Consider a star in the sky at a distance of 10 parsecs (32.6 light-years) that lies in the ecliptic, that is, it lies in the same plane as that of Earth's orbit around the Sun. Sketch the path that this nearby star traces on the sky relative to the background stars over the course of the year, as seen from Earth. Explain your reasoning, and indicate on your diagram its dimensions in arcseconds.

36.b Imagine drawing an axis perpendicular to the plane of the ecliptic; this will point to the ecliptic pole. Consider a star, also 10 parsecs away, that lies in the direction of the ecliptic pole. Again, sketch the path this star appears to trace over the course of the year relative to distant stars behind it, and indicate the dimensions of your diagram.

36.c Now imagine that the star in part **a** has a "proper motion," that is, it is moving through space in a direction perpendicular to the plane of the ecliptic. Describe qualitatively how the motion you described in part **a** is modified.

37. Really small angles and distant stars

Astronomers are fond of measuring really small angles. A European spacecraft called Gaia is measuring the parallax of stars to unprecedented accuracy.

37.a For the brightest stars, Gaia will measure parallaxes to an accuracy of 10 micro-arcseconds. One micro-arcsecond is 10^{-6} of an arcsecond. To get a feel for what a tiny angle this is, calculate the diameter of an object in Los Angeles, seen from New York (4,000 kilometers away) that would subtend an angle of 10 micro-arcseconds. Here we'll ignore the fact that the curvature of the Earth means that you cannot get a direct view of Los Angeles from New York! Express your answer in microns (10^{-6} meter).

37.b If the uncertainty in a given parallax measurement is 10 micro-arcseconds, then the smallest parallaxes that Gaia can robustly measure (i.e., with 20% accuracy) is 5 times that, or 50 micro-arcseconds. What is the distance of a star whose parallax is 50 micro-arcseconds?

38. Brightness, distance, and luminosity

This problem will exercise your knowledge of the relationship between temperature, size, and luminosity of a star, and between that luminosity, the star's distance, and its brightness.

38.a Consider a star with a brightness of 6×10^{-11} erg/sec/cm^2. For those of you who are familiar with stellar magnitudes, this corresponds to about 14th magnitude, much fainter than one can see with the naked eye. This star is a cool main sequence star, with a mass 1/10 that of the Sun. It has a surface temperature of 2,900 K (half that of the Sun). It has a radius about 1/10 of the Sun's radius (a typical number for low-mass stars on the main sequence), and it radiates like a blackbody. Calculate its luminosity. (*Hint: Scale to the Sun; you can use the fact that the Sun's luminosity is 4×10^{33} erg/sec.*) How far away is the star in light-years? What is its parallax (in arcsec)?

38.b Now consider a star with the same brightness, but this one is a red giant, with a surface temperature of 4,500 K and a radius of 1 AU (!). You are observing in a direction well out of the plane of the Milky Way, so there is no interstellar dust to worry about. How far away is this star? Compare with the dimensions of the Milky Way galaxy, and comment.

39. Comparing stars

Consider two stars, each radiating like blackbodies. One, a main sequence O star, has a surface temperature of 30,000 K. The second, a red giant, has a surface temperature of 3,000 K and a luminosity 16 times larger than that of the O star.

39.a Given its surface temperature, what is the peak wavelength at which the O star emits? Express your answer in microns (1 micron = 10^{-6} meters).

39.b What is the ratio of the radius of the red giant to that of the O star?

39.c The two stars appear equally bright in the sky. The distance to the O star is 1,000 light-years. How far away is the red giant? Express your answer in light-years.

40. Hot and radiant

This problem concerns the luminosities, sizes, and temperatures of stars and planets.

40.a Jupiter has an average surface temperature of about 200 K, while the Sun's surface temperature is about 6,000 K. The Sun's radius is 700,000 km; Jupiter's radius is roughly 1/10 of that. What is the ratio of the blackbody luminosity of the Sun to that of Jupiter? At what wavelength does Jupiter's emitted radiation peak? Express your answer in microns.

40.b A hot white dwarf star has a diameter of about 14,000 km and a surface temperature of 40,000 K. What is the ratio of its luminosity to that of the Sun? At what wavelength does the white dwarf's emitted radiation peak (in microns)? What color is it?

40.c Consider a red giant star that has a diameter 80 times that of the Sun but a surface temperature of only 3,000 K. What is the ratio of its luminosity to that of the Sun? At what wavelength does the red giant's emitted radiation peak? What color is it?

41. A white dwarf star

Consider a hot white dwarf star that has the same luminosity as does the Sun. It can be seen with a moderate-size telescope; it is 10^{16} times less bright than is the Sun. From its spectrum, its surface temperature is measured to be a blazing 60,000 K. In what follows, it will be useful to scale from the known properties of the Sun.

41.a How far away is the white dwarf? Express your answer in light-years.

41.b What is the radius of the white dwarf? Express your answer in kilometers. You should assume that the white dwarf and the Sun both radiate like a blackbody.

42. Orbiting a white dwarf

This problem will give you a sense of just how massive a white dwarf is relative to Earth, even though they have similar radii.

42.a The space shuttle, in low-Earth orbit (i.e., in a orbit just above Earth's surface), takes roughly 90 minutes to orbit Earth. Now consider a spaceship in orbit just above the surface of a white dwarf of the same radius as Earth, but with a mass equal to that of the Sun. What is the orbital period?

42.b Would astronauts in such a spaceship feel weightless? Explain your answer.

43. Hydrogen absorbs

This is a more challenging problem.

A great deal of information is encoded in the spectrum of a star. Here you will use the spectrum of an F star to learn about the physics of hydrogen atoms.

43.a Figure 2 shows the spectrum of an F star (this is real data!). Essentially all the absorption lines (the broad dips) you see here are due to hydrogen. Measure the central wavelengths of the lines as well as you can. The wavelengths of the lines are given by a formula you may have seen in high school physics or chemistry:

$$\frac{1}{\lambda} = R \left(\frac{1}{n^2} - \frac{1}{m^2} \right),$$

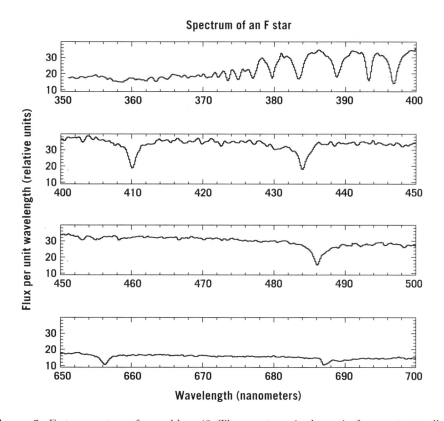

Spectrum of an F star

Figure 2. F star spectrum for problem 43. The spectrum is shown in four parts, to allow you to accurately identify the position of absorption lines (the dips in the spectrum). Data are from the Sloan Digital Sky Survey.

where R is a constant, and n and m are small integers. Identify as many of these lines as you can (you should find 10 or more). This means, first, find the value of R, and then figure out the values of n and m for each one. You might indeed find a calculator useful for this exercise! *Hints: The value of n is the same for all the lines you see here, and is a very small integer. The spectrum does include the so-called α line, for which $m = n + 1$. Also, there are a few lines that are not due to hydrogen and thus do not follow the above equation. A stretch of the spectrum from 500 to 650 nm is not shown in the figure: no strong absorption lines lie in that region.*

43.b At what wavelength does the flux of the star peak? Make sure you use the overall peak of the spectrum, and don't get confused by the local peaks between absorption lines. Use the peak wavelength to calculate the surface temperature of the star.

7–8

THE LIVES AND DEATHS OF STARS

44. The shining Sun

In this problem, you are going to see just how quickly the Sun uses up the nuclear fuel in its core.

44.a Four hydrogen atoms (total mass $= 6.693 \times 10^{-27}$ kilograms) fuse to produce one helium atom (mass $= 6.645 \times 10^{-27}$ kilograms). If 1 gram of hydrogen is burned in the Sun into helium, how much energy will be produced?

44.b The Sun has a luminosity of 4×10^{26} Joules per second. It has a mass of 2×10^{30} kilograms. Three quarters of the Sun's mass is hydrogen. Make a rough estimate of the maximum time the Sun could shine at its present luminosity by burning hydrogen into helium. How does this estimate compare with its actual main sequence lifetime? Explain any discrepancies you might find.

45. Thermonuclear fusion and the Heisenberg uncertainty principle

This is a more challenging problem.

In this problem, we will explore the physics of the thermonuclear reactions in the cores of stars, whereby protons fuse to create helium nuclei. *Welcome to the Universe* notes that if two protons come close enough to each other, the strong force will overcome the electrostatic repulsion between them and cause them to fuse. Our goal will be to calculate how hot the interior of the star has to be to get them close enough, and as we'll see, it is a bit tricky to figure out what "close enough" means.

45.a First we determine the relationship between the temperature T of a gas of protons and how close two protons in that gas can approach each other. Consider two protons racing toward each other, each with initial kinetic energy $E = \frac{3}{2}kT$. Here k is Boltzmann's constant, given in the Useful Numbers and Equations. Their mutual repulsion (due to the fact that they are both positively charged) slows them down, until they come to a stop before whizzing away from each other again: at the point of closest approach, all that kinetic energy has been turned into *electrostatic potential energy*. The expression for the potential energy U of two protons of charge e separated by a distance r is

$$U = q\frac{e^2}{r},$$

where $q \approx 9 \times 10^9$ Joule meter Coulomb^{-2} is called the *Coulomb force constant*. The charge of a proton is $e \approx \frac{1}{6} \times 10^{-18}$ Coulombs.

Suppose that the strong force takes over when two protons are touching directly (i.e., when $r = 10^{-15}$ m). Equate the initial kinetic energy to the final potential energy at this distance, and solve for the temperature needed to get the two protons this close together. Compare to the temperature at the center of the Sun (15 million K) and comment: by your calculation, is it hot enough in the center of the Sun to allow two protons to fuse?

45.b You should have found in part **a** that the temperature needed to get two protons to fuse is *much* larger than the true temperature at the core of the Sun. Yet we know that thermonuclear fusion does take place in the Sun, suggesting that we're missing something in our understanding. That something is quantum mechanics. One of the tenets of quantum mechanics is that an object's exact position, and its exact speed, cannot both be known simultaneously (this is known as the *Heisenberg uncertainty principle*). In particular, the position of a proton of mass m and speed v can be pinned down to an accuracy only of

$$\lambda = \frac{h}{mv},$$

where $h \approx \frac{2}{3} \times 10^{-33}$ Joule sec is known as *Planck's constant* (the distance λ in the equation above is referred to the *de Broglie wavelength* of the proton). First, show that λ has units of distance. Second, use the relationship between kinetic energy and the temperature of a gas to express v, and therefore λ, in terms of temperature (algebra only at this point; we'll plug in numbers below). Third, redo the calculation in part **a**, substituting the de Broglie wavelength for the distance between two protons when the strong force takes over, and solve for temperature. Express your answer algebraically, in terms of q, e, m, h, and k. Finally, plug in numbers. Is the resulting temperature in better agreement with the true temperature in the core of the Sun?

46. Properties of white dwarfs

This is a more challenging problem.

In this problem, we will go directly from observed quantities to the density of a white dwarf star.

46.a Sirius B, the white dwarf companion to Sirius A (the brightest star in the night sky), has a parallax due to the Earth's orbit around the Sun of 0.38 arc-seconds. What is its distance? Give the result in light-years.

46.b If it weren't so close to its very bright companion, Sirius B could be seen with a good pair of binoculars. Its measured brightness is 10^{-13} that of the Sun (it is, after all, much more distant than the Sun!). Calculate the ratio of the luminosity of Sirius B to that of the Sun.

46.c Sirius B has a surface temperature of $\approx 25,000$ K (as measured from the spectrum). Assuming that it radiates like a blackbody (indeed it does), and armed with your knowledge of the star's luminosity from part **b**, calculate the radius of Sirius B. *Hint: If you find yourself using the Stefan-Boltzmann constant in your calculation, you will get the right answer, but you are working too hard. Instead, you should scale from your knowledge of the radius and surface temperature of the Sun.*

46.d Because Sirius B and Sirius A are in orbit around each other, the mass of each can be determined using Newton's law of gravity. It turns out that Sirius B has a mass very close to that of the Sun. What is its mean density? You may find amusing the following quote from Sir Arthur Eddington, one of the top astrophysicists of his day:

> The message of the Companion of Sirius, when decoded, ran: "I am composed of material 3,000 times denser than anything you have ever come across; a ton of my material would be a little nugget that you could put in a matchbox." What reply can one make to such a message? The reply which most of us made in 1914 was—"Shut up. Don't talk nonsense."

Check Sir Arthur by calculating how large a 1-ton (1,000 kg) nugget of white dwarf material would actually be. What are the dimensions of a cube of white dwarf material of this mass? Would it fit into a matchbox?

47. Squeezing into a white dwarf

Under the tremendous densities in the interior of a white dwarf $(2 \times 10^9 \, \text{kg/m}^3)$, the individual atoms cannot hold on to their electrons; the nuclei are squeezed together while the electrons wander between them, unassociated with any nucleus. Consider the white dwarf to be made entirely of carbon nuclei (12 times the mass of hydrogen); the contribution to the white dwarf mass from the electrons is negligible. Calculate the mean distance between nuclei in a white dwarf; express your result in Ångstroms (1 Ångstrom $= 10^{-10}$ meters). Compare with the typical distances between nuclei in solid carbon under normal conditions, ~ 1 Ångstrom. *Hint: We are looking for an approximate answer here.*

48. Flashing in the night

Pulsars are objects that are observed to flash up to 600 times per second. A pulsar is a spinning star with a bright spot near its magnetic pole, and we're seeing the bright spot every time the star completes a rotation. Thus the rotational period of such a star is 1/600 second. In this problem, we will try to understand the physical nature of a pulsar.

48.a Consider whether pulsars could be white dwarfs: A typical white dwarf of 1 solar mass has a radius about equal to that of Earth (6,400 km). What would be the speed (in meters per second) of a point on the rotational equator of such a white dwarf, if it were spinning with a period of 1/600 second? Are there any laws of physics that would prevent such speeds?

48.b Consider whether pulsars could be neutron stars: A typical neutron star of 2 solar masses has a radius of 10 km. Again, calculate the speed of a point on the rotational equator of such a neutron star, spinning with a period of 1/600 sec. Is this speed physically possible?

48.c The speed you have just calculated in part **b** is quite high. Stars are held together by gravity. Consider a chunk of material on the surface of the 2 solar-mass neutron star: the gravitational force on it must be large enough to supply the centripetal acceleration required to move the object in a circle. Let's calculate—is the gravity of a neutron star adequate to keep an object on its equator fixed to the star? If not, the neutron star would fly apart. *Hint: Compare the gravitational acceleration on the surface of the neutron star to the centripetal acceleration; which must be greater so that the neutron star will not fly apart?*

49. Life on a neutron star

The extreme gravity on the surface of a neutron star makes it an environment completely alien to our everyday experience.

49.a Consider what life on the surface of a neutron star might be like. For definiteness, we will consider a 2-solar-mass neutron star with radius 10 kilometers. In a gravitational field with acceleration g, the energy you expend to move an object of mass m a height h is mgh. Calculate, in Joules, the energy you would have to exert to lift 1 gram a distance of 1 centimeter. *Hint: To do so, you will first need to calculate the acceleration of gravity on the surface of the star.* Compare with your daily energy intake from food (2,000 calories; 1 calorie of food $= 4,000$ Joules). How many days' worth of food would you have to eat to have the energy to lift this mass?

49.b Suppose you were falling feet-first toward the neutron star. You are weightless as you fall. But bad things happen to you even before you hit the ground. The gravitational acceleration is strong, as you calculated in part **a**, but the difference in gravitational acceleration between your head and feet is also huge. Calculate this difference in units of meters sec^{-2}, just before you hit the surface. This difference (which is called the *tidal force*) is so great that your body would be torn to shreds. *Hint: The best way to do this problem is to remember that the distance between your head and feet is much smaller than the radius of the neutron star; when taking the difference requested in the problem, take advantage of this fact to greatly simplify the problem. Do this problem algebraically first before plugging in numbers. Those of you who know calculus will recognize this process as calculating a differential.*

50. Distance to a supernova

One of the great challenges in astronomy is determining how far away the objects we see in the sky are. Here you will estimate the distance to a particularly famous supernova remnant, one whose explosion was seen here on Earth almost 1,000 years ago.

50.a The Crab Nebula, a supernova remnant, is roughly spherical and is expanding. It is depicted in figure 8.4 of *Welcome to the Universe*. The angle its diameter subtends on the sky increases at a rate of 0.23 arcseconds per year. Spectra taken of the glowing gases indicate, via the Doppler shift, that the gas is also expanding along the line of sight at 1,200 km sec^{-1} relative to the center. From this, deduce an approximate distance to the Crab Nebula in light-years. *Hint: Assume that the explosion is isotropic (i.e., the same in all directions). That is, the line-of-sight expansion speed is the same as that in the plane of the sky. If the expansion rate relative to the center is 1,200 km/s, ask yourself what the speed is from one side of the supernova remnant to the other.*

50.b The angular size of the Crab Nebula is currently 5 ± 1.5 arcminutes in diameter (the error reflects the fact that the nebula is not exactly spherical). Assuming that it has been expanding at a constant rate, calculate roughly the year in which the light from the initial supernova event reached Earth. Be sure to include the effects of the uncertainty in the angular size. Compare your result with the date at which the Chinese observed this supernova, 1054 AD. Do these dates agree within the precision of your calculation?

50.c You are given the velocity of the gas in part **a**. The mass of the expanding gas is about 20 solar masses. Calculate the kinetic energy of the explosion; express your result in Joules. The star that exploded was probably an O star, with a main sequence luminosity 10^3 times that of the Sun. Calculate how long the O star has to shine to generate as much energy as is currently present in the form of kinetic energy of the expanding gases.

51. Supernovae are energetic!

The explosion of a supernova is a tremendously powerful event, releasing as much visible light energy in a month as the Sun will emit in 100 million years.

51.a Calculate the total amount of energy emitted in the form of light by a supernova. Express your answer in Joules.

51.b What is the average luminosity of the supernova? Express your answer in units of solar luminosities.

51.c At what distance would the supernova appear as bright in the sky as the full Moon? Express your answer in light-years. The full Moon has a brightness 4×10^5 fainter than the Sun.

52. Supernovae are dangerous!

The explosion of a nearby supernova could have a catastrophic effect on life on Earth. In the previous problem, you considered a supernova close enough to appear as bright as the full Moon. While this would be a spectacular sight in the sky, it would appear harmless. Unfortunately, in addition to the visible-light photons, there are many much higher energy photons and cosmic rays that the supernova would emit, which could do life-threatening damage to Earth's atmosphere, even if the supernova were 100 light-years (30 parsecs) from Earth.

There is roughly one star per cubic parsec in the Milky Way. Roughly one supernova explodes per 100 years among the 10^{11} stars in the Milky Way. Roughly how long will it be before the next supernova explodes close enough to Earth (within 30 parsecs) to significantly affect Earth's atmosphere? Based on your answer, is this a major concern for our future?

53. Neutrinos coursing through us

In 1987, the astronomical world was electrified by the news of a supernova exploding in the Large Magellanic Cloud, a dwarf galaxy companion to the Milky Way, at a distance of 150,000 light-years. It was the nearest supernova to have gone off in 400 years and was studied in great detail. Its luminosity was enormous; the explosion released as much visible light energy in a few weeks as the Sun will emit in 10^8 years. It was easily visible to the naked eye from the Southern Hemisphere. However, theories of the mechanisms taking place in the supernovae predict that the visible light represents only a small fraction of the energy emitted by the supernova, most of which is emitted in the form of neutrinos. When the supernova was discovered, John Bahcall of the Institute for Advanced Study in Princeton and his colleagues did the calculation that follows, asking whether any of the neutrinos emitted from the supernova should have been detected here on Earth.

53.a The energy of the supernova explosion is mostly driven by the gravitational collapse of the core of the star when that core becomes iron, thus turning off the central furnace. The gravitational potential energy of that core of mass M and radius r is approximately $\frac{GM^2}{r}$, where G is Newton's gravitational constant. That energy is released during the collapse in the span of a second or two! The core of the star has a mass equal to that of the Sun and has a radius comparable to a neutron star, 10 kilometers. Calculate the total energy emitted from the supernova in units of Joules.

53.b We said that the energy you have just calculated is emitted mostly in the form of neutrinos; the visible light from the supernova represents only about 0.01% of the total energy of the explosion. Each neutrino has an energy of roughly 1.5×10^{-12} Joules. Calculate how many neutrinos are emitted by the supernova. (This is an easy calculation, but it will give you a very large number!)

53.c Scientists have built a number of detectors of neutrinos (to look for neutrinos from the Sun, another interesting subject). One of the largest is called Kamiokande in Japan. In 1987, it consisted of a cube of water roughly 10 meters on a side (it has since been expanded). Calculate how many neutrinos from the supernova would have passed through the cube of water. *Hint: The inverse square law holds for neutrinos just as it does for light. Consider the area of the detector, and remember that the neutrinos are sent out in all directions from the supernova; at the time that the neutrinos hit Kamiokande, they are spread out over an enormous sphere centered on the supernova.*

53.d Neutrinos are ghostly particles; they are very difficult (but not completely impossible) to detect. A typical neutrino can pass through the entire Earth unimpeded, without anything happening to it. However, it turns out that roughly 1 in 5×10^{15} (i.e., 1 in 5 quadrillion!) of the neutrinos passing through Kamiokande will interact with the water in the detector, and thereby be detected. Calculate how many neutrinos should have been detected by Kamiokande.

54. A really big explosion

Astronomers have long been intrigued by gamma-ray bursts, which, as the name implies, are sudden bursts of high-energy photons seen to suddenly go off randomly in the sky. Still an area of intensive research, it is now thought that at least some of these gamma-ray bursts may represent the final death throes of a massive star collapsing into a black hole. Gamma-ray bursts are quite rare: a NASA satellite dedicated to their discovery called the Swift Gamma-Ray Burst Mission detected fewer than 100 bursts per year. But

March 19, 2008, was a special day for the Swift satellite; it discovered no fewer than five gamma-ray bursts, one of which (called GRB 080319B) was the most powerful such burst ever detected to that time. Indeed, this object was briefly the most-luminous known object in the universe. In this problem, you will calculate its properties and learn just how awesome this discovery was. You will be doing the same calculations that professional astronomers all over the world did when they first heard the news.

54.a The Burst Alert Telescope (BAT) on Swift detected about 70,000 photons per second from GRB 080319B; the burst lasted about 30 seconds. BAT has an effective collecting area of 500 cm^2, once all its efficiencies are taken into account. These gamma-ray photons each have an energy of about 100 kiloelectron volts (keV). For reference, 1 keV is about 1.6×10^{-16} Joules. Calculate the gamma-ray brightness of GRB 080319B in standard MKS units (Joules/second/meter^2).

54.b Gamma-ray bursts often also are accompanied by emission from visible-light photons. The visible-light emission from GRB 080319B was bright enough to (barely) be seen by the naked eye, at a brightness of about 1×10^{-10} Joules/sec/m^2 (corresponding to about 6th magnitude). The visible-light burst also lasted about 30 seconds. What is the ratio of the brightness in gamma-ray photons to that in visible light?

54.c When astronomers measured the spectrum of the visible light from this GRB, they were able to determine its *redshift*. As chapter 14 of *Welcome to the Universe* describes, this allows a determination of its distance. The redshift is 0.94, corresponding a current distance of an astonishing 11 billion light-years. It is truly awesome that an object so far away was (briefly) visible to the naked eye! Calculate the total amount of energy (in Joules) emitted by this object. Do the calculation separately for the gamma-ray energy and the visible-light energy. *Hint: Remember that luminosity is energy per unit time, with units of Joules per second. So calculate the luminosity, and multiply by the length of time the burst lasted. Because of the expansion of the universe, and the high redshift, the energy per photon received by us is smaller than the emitted energy by a factor of $(1 + \text{redshift}) \approx 2$. This factor holds both for visible light and gamma-ray photons. So you will need to multiply your result by a factor of 2.*

54.d To get a sense of just how large the numbers you calculated in part
c are, imagine that all the gamma-ray energy was created by the
conversion of mass into energy via Einstein's equation $E = mc^2$. How
much mass does this energy correspond to? Express your answer in
units of the mass of the Sun.

54.e Even though you have found that the visible-light energy is less than
that of the gamma rays, it is enormous in its own right. How long
would the Sun have to shine at its current rate to emit as much visible
light energy as the gamma-ray burst did? Express your answer in
years, and compare with the 10-billion-year lifetime of the Sun.

55. Kaboom!

On December 27, 2004, gamma-ray detectors on a variety of satellites detected something amazing: the most powerful blast of gamma
rays ever detected. For 0.1 seconds, the detectors received a flux of
10^{11} photons s^{-1} m^{-2}, in the form of gamma rays of wavelength 7×10^{-12} m.
It was determined that these were coming from an object 50,000 light-years
away, on the other side of the Milky Way Galaxy.

55.a Calculate the energy of a single gamma-ray photon; express your
energy in Joules. With this in hand, convert the flux to a brightness,
in units of Joules $\sec^{-1}$ m^{-2}.

55.b Calculate the luminosity of this object, knowing its brightness and
distance. Express this luminosity in units of Joules per second (i.e.,
watts). Knowing that the source maintained this luminosity for only
0.1 seconds, calculate the total amount of energy it generated (in
Joules), and calculate how long it would take the Sun to generate
the same amount of energy.

55.c The Swift satellite (see problem 54) was one of those that detected
this emission; it could have detected a gamma-ray flux of only
10^4 photons $\sec^{-1}$ m^{-2}. How far away could such an object be and still
be detectable by Swift? Express your result in light-years.

56. Compact star

Ordinary matter that you encounter every day in solid objects is made
entirely of atoms. Each atom is roughly 10^{-10} meters in diameter; this
distance separates the centers of atoms in ordinary solid material. However,

most of an atom is empty space, with the nucleus (which holds nearly all the mass of the atom) having a diameter of 10^{-15} meters.

In a neutron star, all this empty space is squeezed out, so that the nuclei are touching one another. That is, the distance between the centers of nuclei in a neutron star is 10^{-15} meters. Calculate the ratio of the density of a neutron star to that of an ordinary solid object in your home. Remember that the density of an object is its mass divided by its volume. This calculation should be done to one significant figure (and therefore you need not worry about the exact geometrical configuration of the atoms or nuclei). *Hint: If you find yourself having to plug in the mass of an atom in your calculations, you are doing this problem the hard way.*

57. Orbiting a neutron star

In its orbit around Sun, the Earth travels at a speed of 30 kilometers per second. Now consider a planet in a circular orbit around a neutron star of the same mass as that of the Sun. The radius of the orbit is 9 AU. Calculate its speed in units of kilometers per second.

58. The Hertzsprung-Russell diagram

The Hertzsprung-Russell (H-R) diagram is a basic tool that astronomers use to understand the properties of stars. Write an essay explaining what the H-R diagram is, sketch it (be sure to label your axes!), and explain how the placement of stars on the diagram relates to the masses, temperatures, radii, lifetimes, and evolutionary stages of stars. Your explanation should clearly explain the nature of main sequence, red giant, and white dwarf stars, and the evolutionary relationship between the three categories of stars. As part of your essay, discuss thermonuclear fusion and its role in different types of stars.

9

WHY PLUTO IS NOT
A PLANET

59. A rival to Pluto?

This is a more challenging problem.

The discovery of the minor planet Eris threw the astronomical community into a tizzy and made international headlines; it is almost as large as Pluto (as we'll see) and brings up interesting questions about what the definition of a planet is.

You have been named to head a crack team of astrophysicists to calculate some of the properties of Eris. In this problem, you will be doing groundbreaking research, and indeed are using the same tools that professional astronomers use to understand the properties of this object. Throughout, assume that Eris is spherical and is observed at opposition (i.e., Earth lies on the straight line connecting the Sun and Eris).

59.a You get a call from Mike Brown of Caltech, the leader of the team that discovered Eris and other Pluto-sized objects in the Kuiper Belt. "I need your help to do some important calculations about my new-found planet," he explains. "I need to figure out how far away it is." In 5 hours, Eris is observed to move 7.5 arcseconds relative to the background stars as seen from Earth (see http://www.gps.caltech.edu/~mbrown/planetlila/lila.gif). This observation is made when Eris is in opposition (i.e., opposite the Sun as seen from Earth). Because (as we'll see) it is much farther from the

Sun than Earth, it is moving quite a bit slower around the Sun than Earth does, so this apparent motion on the sky is essentially entirely parallax due to Earth's motion. First, calculate the speed that Earth goes around the Sun, in kilometers/second. Then use this information and the small-angle formula to calculate the distance from Earth to Eris. Express your result in AU. Compare with the semi-major axis of Pluto's orbit (which you'll need to look up).

59.b We're next going to calculate how big Eris is from its apparent brightness. We'll have to do a few calculations along the way. Eris shines in two ways: from its reflected light from the Sun (which will be mostly visible light), and from its blackbody radiation from absorbed sunlight (which will mostly come out as infrared light); it will be important in the calculations that follow to keep in mind which of two you're dealing with. The albedo of Eris (i.e., the fraction of the sunlight incident on Eris that is reflected) is very high, about 96%. This suggests that Eris is covered by a layer of shiny ice; spectroscopy tells us that the ice is composed of frozen methane, CH_4.

We are going to start by deriving an expression for the brightness of Eris that depends on its distance from the Sun d, and its radius r. First, calculate the amount of sunlight reflected by Eris per unit time (i.e., its luminosity in reflected light); express your answer in terms of d, r, the albedo A, and the luminosity of the Sun L. Don't plug in numbers yet!

59.c We detect only a tiny fraction of this light reflected by Eris. Calculate the brightness, via the inverse square law, of Eris as perceived here on Earth. Again, don't plug in numbers yet; just do algebra; again your answer will involve d, r, A, and L. *Hint: With Eris so distant from the Sun, the tiny 1 AU separation between Earth and the Sun is negligible. So make the approximation that the distances from the Sun to Eris, and from Eris to Earth, are the same.*

59.d OK, now for some numbers. The measured brightness of Eris is 2.4×10^{-16} Joules meters^{-2} second^{-1}. We know the distance d to Eris and its albedo A, and can look up the luminosity of the Sun L. Use this information to solve for the radius r of Eris. *Hint: Be careful to be consistent with your units!* Compare with the radius of Pluto (1,200 km); you should find a comparable result. As described in chapter 9 of

Welcome to the Universe, this is what led to the controversy about the definition of a planet: if Pluto is considered a planet, then certainly Eris should be as well.

59.e Your phone rings again; it is Ken Sembach, director of the Hubble Space Telescope. "We have observed Eris with the Hubble Telescope, and have found that it has a moon orbiting it!" he says. "I am told that you are an expert in the use of Kepler's third law; we need you to calculate Eris's mass." The moon is named Dysnomia; those of you who are familiar with obscure characters in Greek mythology may get a chuckle out of these names. Observations with Hubble show that Dysnomia makes an almost circular orbit around Eris with a period of 15.8 Earth days. The semi-major axis of the orbit subtends an angle of 0.53 arcsecond as seen from Earth. Calculate the semi-major axis in kilometers, and calculate the mass of Eris in kilograms. Compare with the mass of Pluto (1.3×10^{22} kg). Is Eris more massive?

59.f Alan Stern, a world expert on Pluto, is next on the phone. "Rumor has it you've determined both the mass and radius of Eris. Do you realize how important this is? It means that you can calculate its density, which gives clues as to what it is made of." What is this density (in units of grams per cubic centimeter)? How does it compare with the density of water ice (1.0 g/cm^3), Pluto (1.9 g/cm^3, whose mass is probably dominated by silicate rocks), and Earth (5.5 g/cm^3, whose mass is dominated by iron and nickel in the core)? Comment on the possible composition of Eris.

59.g The phone is ringing off the hook. This time it's Dr. Lisa Storrie-Lombardi, director of the Spitzer Science Center. They operate the Spitzer Space Telescope, a NASA telescope sensitive in the infrared part of the spectrum. "I understand you've learned how to calculate the temperature of bodies in the solar system," she says. "Could you calculate the expected equilibrium temperature of Eris?" (You should ignore the greenhouse effect; Eris is cold enough that it is unlikely to have much atmosphere.) Dr. Storrie-Lombardi goes on. "I need to know at what wavelength the spectrum of blackbody radiation from Eris peaks, so that we know how to configure the SST's instruments to observe it." Answer her question, expressing your answer in microns. Is this indeed in the infrared part of the spectrum?

59.h The phone rings yet again. Charles Bolden, chief NASA administrator, is thinking of sending a spacecraft to Eris. "I've heard about your groundbreaking calculations of the properties of Eris," he says breathlessly. "I need to know whether Eris is likely to have a nitrogen atmosphere." Calculate the escape speed from the surface of Eris. Compare with the speed of a nitrogen molecule (N_2) in its atmosphere, given the surface temperature you calculated above. Can Eris hold onto its nitrogen, or will it escape?

Your phone is now quiet. You collapse exhausted but happy, hopeful that the next phone call will be from the Nobel Prize committee in Stockholm....

60. Another Pluto rival

This problem is closely related to problem 59, but these problems are so much fun, it is worth going through this twice! And the numbers in each case are **not** *made up, but represent real measurements.*

The Kuiper-Belt Object Orcus is one of the biggest asteroids known in the solar system. You have been named to head a crack team of astrophysicists to calculate some of its properties; in doing so, you will learn a few general things about Kuiper Belt Objects (also known as Trans-Neptunian Objects, or TNOs) along the way. In this problem, you are doing groundbreaking research and indeed are using the same tools that professional astronomers use to understand the properties of this object. Throughout, assume that Orcus is spherical, is observed at opposition, and is in a circular orbit around the Sun.

60.a In 1 hour, Orcus is observed to move 3.8 arcseconds relative to the background stars. As you'll show below, it is moving quite a bit slower around the Sun than is Earth, so this apparent motion on the sky is essentially entirely parallax due to the Earth's motion. Use this information to calculate the distance from Earth to Orcus. Express your result in AU. Compare with the semi-major axis of Pluto's orbit.

60.b Calculate the orbital period (in years), and the speed (in km/sec) at which Orcus is orbiting the Sun. Is it indeed traveling much slower than is Earth, as we assumed in part **a**?

60.c Now calculate the orbital period of Neptune (semi-major axis of 30 AU) in years, and then calculate the ratio of the orbital period of Orcus to that of Neptune. You should find that this ratio is very close to the ratio of two small integers; what are the integers? As a result, every few orbits, Neptune and Orcus find themselves at the same position in their orbits relative to one another, and thus they have the same gravitational interaction with each other, which is reinforced with time. When this happens the same way over and over again, it is called a *resonance*, and this particular resonance is one that attracts many asteroids to it. Check out the listing of all the TNOs known to date at http://cfa-www.harvard.edu/iau/lists/TNOs.html, and skim your eye down the semi-major axis column (marked with the symbol "a"). Guesstimate the fraction of known TNOs that are in this resonance with Neptune. Such objects, which share an orbit with Pluto, are known as *Plutinos*.

60.d We're next going to calculate how big Orcus is, from its perceived brightness. We'll have to do a few calculations along the way. Orcus shines in two ways: from its reflected light from the Sun, and from its blackbody radiation from absorbed sunlight; it will be important in the calculations that follow to keep in mind which of two you're dealing with. The albedo of Orcus is 0.23. That is, it reflects only 23% of the light incident on it, so this object is fairly dark.

In fact many of the objects in the outer solar system are even darker, with albedos of 10% or even lower. It is only poorly understood why this is so; it is telling us something about the chemical makeup of the surfaces of these Kuiper Belt Objects. Interestingly, Pluto itself is an exception. It is quite reflective (60%), due to a layer of methane and nitrogen frost on its surface.

We are going to start by deriving an expression for the brightness of Orcus that depends on its distance from the Sun d and its radius r. First, calculate the amount of light reflected by Orcus; express your answer in terms of d, r, the albedo A, and the luminosity of the Sun L. Don't plug in numbers yet.

60.e We detect only a tiny fraction of this light reflected by Orcus. Calculate the brightness, via the inverse square law, of Orcus as perceived here on Earth. Again, don't plug in numbers yet; just do algebra; again your answer will involve d, r, A, and L. *Hint: With Orcus so distant from the Sun, the tiny 1 AU separation between Earth and the Sun is negligible. So make the approximation that the distances from the Sun to Orcus and from Orcus to Earth are the same.*

60.f Now for some numbers. The measured brightness of Orcus is 4×10^{-16} Joules meters^{-2} second^{-1}. We know the distance d to Orcus and its albedo A, and we can look up the luminosity of the Sun L. Use this information to solve for the radius r of Orcus. *Hint: Be careful to be consistent with your units!*

61. Effects of a planet on its parent star

When we considered the orbits of planets around a star like the Sun, we made the assumption that the star stayed perfectly still. That is not exactly true: Newton's third law of motion says that a planet exerts a gravitational pull on its star of the same magnitude (but in the opposite direction) that the star exerts on the planet. This problem is related to problem 30, which also explores the effect of an orbiting planet on its parent star.

We want to calculate the speed of the star itself; we'll see that this leads to a way to infer the existence of the planet. Here we consider the simple case of a star with a single planet going around it.

61.a One consequence of Newton's third law is that *momentum* is conserved. The momentum **p** of an object of mass m moving at a velocity **v** is their product, $\mathbf{p} = m\mathbf{v}$. The sum of the momenta of the planet and star is zero, which is to say that relative to their common center of mass, the momentum of the star is always equal and opposite to that of the planet. Starting from the expression for the speed of a planet in a circular orbit of radius r from a star of mass M_*, $v = \sqrt{\frac{GM_*}{r}}$, derive an algebraic expression for the speed of the star. The mass of the planet is M_p.

61.b Use the expression you've just derived to calculate how fast the Sun is moving due to the gravitational pull of Jupiter (this will require looking up the mass and orbital radius of Jupiter); express your result in meters per second. Compare with the speed of the fastest human runners (remember that the world record for the 100-meter dash is about 10 seconds).

61.c Could we detect a Jupiter-like planet orbiting a distant star? The light from a planet is exceedingly faint, and the technology to see this light directly next to the glare of the much brighter star is only now starting to bear fruit. But we can detect the motion calculated in part **b** by analyzing the Doppler shift in the spectrum of the star, whereby the spectrum shifts first to longer wavelengths (when the star is moving away from us) and then to shorter wavelengths (when the star is moving toward us). The effect is tiny but measurable, and it oscillates back and forth through the orbit. For each star, we can measure the amplitude of the Doppler shift and the period over which it oscillates.

Here then are data for two stars which show this signature of planets. Assume that the orbit of each planet is circular and perpendicular to the plane of the sky, and that the parent star in each case has the mass, radius, surface temperature, and luminosity of the Sun. For each star, determine the mass of the planet (in units of Jupiter masses) that is causing this motion, the semi-major axis of its orbit (in AU), and its surface temperature. In calculating the temperature, you may ignore the albedo effect; after all, we know nothing about the atmospheres or surfaces of these planets!

Show your work in detail for one of the planets; there is no need to show the detailed calculations for the other one.

Star	Amplitude (m/sec)	Period (Earth days)
51 Pegasi	56	4.2
HD 156836	464	360

62. Catastrophic asteroid impacts
Problem suggested by Chris Chyba.
It is thought that early in its history, during the period referred to as the Late Heavy Bombardment, Earth was struck by several asteroids large enough to evaporate all of its oceans. From the point of view of life on early Earth, that would be bad. This problem asks you to calculate how many times early Earth might have been struck by an asteroid too small to evaporate all the oceans, but still large enough to evaporate the upper 200 meters of the ocean's surface. This is potentially important, because the upper 200 meters comprises the *epipelagic*, or sunlit zone of the ocean—the region in which enough sunlight penetrates for photosynthesis to be possible. A sudden boiling and evaporation of this layer could wipe out most photosynthetic organisms on Earth, with grim implications throughout the food chain.

62.a The oceans cover roughly 3/4 of Earth's surface. What is the total mass of water in the upper 200 meters of Earth's oceans? Express your answer in kilograms. For this problem, assume that the oceans have an average temperature of 0° C. For water initially at 0° C, heating a kilogram of water enough for it to boil at Earth's sea-level atmospheric pressure requires 2.5×10^6 Joules of energy. How much energy, in Joules, is required to boil the upper 200 meters of all Earth's oceans?

62.b The kinetic energy of the impact will be converted into a tremendous amount of heat energy, which will have devastating effects on Earth. Models estimate that 25% of the asteroid's kinetic energy will go into evaporation of ocean water. Assuming that the asteroid is traveling at 20 km/sec relative to Earth, what is the asteroid mass required to boil the uppermost 200 meters of the oceans? How about to vaporize the entire ocean, with a mean depth of 3.5 kilometers?

62.c Many asteroids are made of rock, with a density of 3.0 g/cm^3. What are the volumes, V, in cubic meters, of the two asteroids whose mass you calculated in part **b**?

62.d Assume the asteroids to be spherical in shape. What are the radii, in kilometers, of the two asteroids? The lunar cratering record suggests that the total number of asteroids bigger than or equal to a radius r that hit early Earth is proportional to r^{-2}. So, for example, the total number of asteroids with radii larger than 200 meters is one quarter the number of asteroids with radii larger than 100 meters. The lunar cratering record suggests that early Earth was hit by three asteroids big enough to evaporate the oceans. Calculate how many times Earth was hit by asteroids big enough to evaporate the sunlit zone of the oceans. Suppose that these impacts were spaced evenly between 4.4 billion years ago and 3.8 billion years ago (i.e., during the Late Heavy Bombardment). What was the average interval of time between impacts that were big enough to evaporate the sunlit zone?

63. Tearing up planets

This is a more challenging problem.

In this problem, we will consider just how close a moon can get to its parent planet before being shredded by tidal forces. That is, we are going to ask where the difference in acceleration from the middle to the surface of the moon due to the planet is larger than the gravitational attraction of the moon itself, holding it together.

Consider a planet with radius R_P. Around it is a moon of radius R_m, in an orbit of radius r (i.e., the distance between the centers of the planet and moon). We want to determine the value of r within which the moon's self-gravity is overcome by the tidal force from the planet. Assume that the planet and moon have the same uniform density ρ, and that the planet is much larger than the moon. The distance between them is of course larger than either of them. Thus $r > R_p \gg R_m$.

63.a Consider a rock at the surface of the moon closest to the planet, and another rock at the center of the moon. Calculate the difference in acceleration of these two rocks due to the planet. Be sure to use the approximation that $R_m \ll r$ (i.e., R_m is much less than r). Argue that this differential acceleration tends to pull the two rocks apart.

63.b Next, calculate the acceleration that holds the two rocks together due to the self-gravity of the moon.

63.c Use these results to find a condition on the distance r at which the tidal acceleration overcomes the self-gravity. If the moon approaches closer than this distance, it will be torn apart by the tides induced in it by the parent planet.

63.d Notice that your result does not depend on ρ or R_m. You have derived the so-called Roche limit. In fact, one can do a more careful job (as Édouard Roche himself did), considering the moon as a fluid and asking when it will split into two due to tidal stretching; when one does this, one finds $r_{\text{Roche}} = 2.44 R_p$. Look up a table of the moons of the planets in the solar system (the NASA website http://solarsystem.nasa.gov/planets/ is a good resource). Do any of the moons of the planets fall within the Roche limits of the planet? Comment on your answer.

10

THE SEARCH FOR LIFE
IN THE GALAXY

64. Planetary orbits and temperatures

Imagine a star with mass four times that of the Sun, a radius equal to that of the Sun, and a surface temperature of 12,000 K. It has two planets in circular orbits around it: the first, Zaphod, has an orbital radius of 1 AU, while the other, Prefect, has an orbital radius of 4 AU. *Hint: In these questions, you may scale from what you know about Earth's orbit around the Sun.*

64.a What is the orbital period of Prefect? Express your result in years.

64.b Assume each planet has no greenhouse effect and zero albedo, and radiates as a blackbody. What is the ratio of the equilibrium temperature of Zaphod to that of Prefect?

64.c Zaphod and Prefect have identical masses, but Zaphod has half the radius of Prefect. What is the ratio of their densities?

65. Water on other planets?

In what is perhaps one of the most important astronomical discoveries of the past decades, planets have been discovered orbiting other stars. One technique to find such planets is by observing the reflex motion of their parent star (see problem 61). In this problem, you will derive the surface temperature of six of the thousands of planets thus discovered. Observations of the variation in Doppler shift of the parent star leads to determination of the period and eccentricity of the planetary orbit. The planets you should consider, and their period and eccentricity, are

Name of planet	Period (days)	Eccentricity	Comment
51 Pegasi	4.23	0.01	First extrasolar planet discovered
HD 209458	3.525	0.11	Regularly eclipses its parent star
55 Cancri b	14.65	0.02	This and the next two have the same parent star
55 Cancri c	44.28	0.34	
55 Cancri d	5,360	0.16	
HD 142415	388	0.5	

All these planets have been found around main sequence G stars (i.e., stars with surface temperature, radius, and mass essentially identical to those of the Sun).

65.a For each of the planets, calculate the semi-major axis of the orbit, and also the distance of closest and farthest approach to the parent star.

65.b For each of the planets, calculate the equilibrium temperature on their surface, and for those with eccentricity of more than 10%, calculate the range of surface temperatures from closest to farthest approach to their parent star. For simplicity, assume no greenhouse effect and assume an albedo of zero (i.e., they absorb all the sunlight incident on them); after all, we know next to nothing about their atmospheres!

65.c Is it possible that liquid water can exist on the surface of any of these planets, assuming that they have an atmospheric pressure similar to that on the surface of Earth? Discuss the effect of the eccentricity of the orbit on the possibility of liquid water and on any life that might exist on the planets.

66. Oceans in the solar system

Problem suggested by Chris Chyba.

Earth is not the only body in the solar system that has, or has had, oceans!

66.a The mean depth of the Earth's oceans is 3.5 kilometers, and oceans cover $\frac{3}{4}$ of Earth's surface. What is the mass of Earth's oceans in kilograms?

66.b Mars has a radius of 3,400 km, about half that of Earth. There is an ongoing controversy over whether early Mars had oceans on its surface—oceans that have since been lost. Even if not, there is strong evidence that a lot of water flowed on Mars' surface in the past, and planetary scientists are very interested in how much water must have once been present. Some scientists have estimated that Mars had a quantity of water that, if distributed evenly over the Martian surface, would have resulted in an ocean 1 km deep. What is the ratio of the mass of this putative early Martian ocean to that of Earth's oceans?

66.c Europa is a moon of Jupiter with radius 1,560 km—a bit smaller than Earth's Moon. There is strong evidence that Europa has a global liquid water ocean about 90 km deep beneath an ice cover 10 km deep. Approximate this amount of water as 100 km of liquid water distributed evenly over a sphere 1,460 km in radius. Compare the mass of the putative Europan ocean to that of Earth's ocean by taking the ratio of the two.

67. Could photosynthetic life survive in Europa's ocean?

This problem was suggested by Chris Chyba.

Europa is one of Jupiter's four "Galilean" moons. There are compelling reasons to believe that it has an ocean of liquid water beneath its cracked icy crust. For that reason, it has been the subject of speculation about the prospects for life for more than a quarter century. This problem asks you to recapitulate some of the early plausibility calculations for the prospects of life on that world.

67.a Europa, like Jupiter, is 5.2 AU from the Sun. What is the solar flux (watts per square meter; i.e., the brightness of the Sun) at Europa? There are microorganisms that live at the ice-seawater interface in Antarctica on Earth that are able to photosynthesize at light levels of 0.1 watt per square meter. Could such an organism photosynthesize at Europa's distance from the Sun?

67.b Europa has a radius of 1,600 kilometers. Assume that at any given time, one 10-millionth of the area of its ice crust is cracked, and that these open cracks expose liquid water directly to space. (While spacecraft images show the surface of Europa to be completely covered with cracks, the vast majority of those have refrozen and thus do not expose liquid water to space.) These cracks are important; light that falls directly on the ice is either reflected or absorbed, and never reaches the underlying liquid water. What is the total area (in square kilometers) of liquid water exposed to space at any time? What is the total power (watts) of sunlight reaching Europa's ocean via the cracks? Remember that only about half of Europa is facing the Sun at any given time.

67.c Assuming that they have a source of carbon (typically CO_2), microorganisms on Earth use the photosynthetic energy they absorb to synthesize organic molecules; with each Joule of energy incident on the microorganism, they create roughly 2×10^{-10} kilograms of organic carbon biomass. Assume that every last photon of light reaching Europa's ocean is harvested by such microorganisms, and that there is plenty of carbon available. Over the course of 1 year, how much biomass could be produced on Europa by microorganisms? A typical bacterium on Earth contains about 10^{-15} kg of carbon. How many microorganisms does this biomass represent?

67.d Many microorganisms are known to thrive deep under the surface of Earth. These so-called extremophiles derive their energy from the interior heat of Earth, and have a very slow metabolism, surviving for perhaps 1,000 years. We might guess that the metabolism of microorganisms on Europa is similar, with an average lifetime of 1,000 years. Assume that the population is in a steady state, in which the number of microorganisms being born matches that dying. What is the total mass, then, of microorganisms on Europa? Earth's steady-state biomass is about 1×10^{15} kilograms. What is the resulting ratio of Europa's biomass to Earth's biomass?

68. An essay on liquid water

Liquid water is essential for life as we know it. NASA likes to say, "follow the water!" in the search for life in the universe. Describe where in the solar system liquid water has been found, and under what conditions it might exist. Discuss the equilibrium temperatures of objects in the solar system and how they relate to liquid water; be sure to describe the importance of the greenhouse effect, and how and why it differs on Earth, Venus, and Mars. Your essay should be three or four paragraphs long.

PART
II

GALAXIES

11–13

THE MILKY WAY AND THE UNIVERSE OF GALAXIES

69. How many stars are there?

In this problem, we will calculate the number of stars in the observable universe.

69.a The Milky Way Galaxy in which we live is shaped like an enormous circular flattened disk. It has a radius of 50,000 light-years and a thickness of about 1,200 light-years. The typical distance between stars is about 4 light-years. Calculate roughly how many stars there are in the Galaxy. *Hint: Think about how much volume that the typical star, plus the empty space around it, occupies.*

69.b The Milky Way is only one of many similar galaxies in the universe. The typical distance between big luminous galaxies like the Milky Way is about 15 million light-years. The visible universe has a current radius of about 45 billion light-years. Estimate roughly how many galaxies there are in the visible universe. *Hint: This calculation is similar to that in part* **a**.

69.c Using the results of parts **a** and **b**, give a rough estimate for the total number of stars in the visible universe. Express your answer both in exponential notation, and for fun, in English units.

Hint: Check out http://mathworld.wolfram.com/LargeNumber.html if you are unfamiliar with the names of really large numbers. Other languages, including Japanese, Chinese, and ancient Sanskrit, have names for such enormous numbers as well.

70. The distance between stars

The nearest star to the Sun, Proxima Centauri, is about 4 light-years away. The Sun's diameter is about 0.01 AU. Calculate the ratio of these two. That is, how many Suns would you have to line up to reach from here to Proxima Centauri? Now imagine scaling the whole problem down: imagine the Sun as a basketball (10 inches in diameter). At that scale, how far away is the nearest star? Express your answer in kilometers.

71. The emptiness of space

In this problem, we will get a sense of just how empty the universe is.

71.a The nearest star to the Sun is about 4 light-years away. This is a typical distance between stars throughout the Milky Way. Assuming crudely that all stars have the same mass as the Sun, and that all the mass of the Milky Way is in stars, calculate the *density* of matter in the Milky Way. Compare with (i.e., take the ratio to) the density of the most extreme laboratory vacuum we can currently produce $(10^{-20}\,\mathrm{g/cm^3})$ and comment. *Hint: You do not need to know the dimensions of the Milky Way to do this problem.*

71.b Now calculate the density of the entire universe, assuming that galaxies all have the same mass $(1.5 \times 10^{11}$ solar masses) and are typically 15 million light-years apart. Here you will be including the contribution from the stars alone and not the dark matter. Express your result in grams per cubic centimeter. What is the radius of a sphere of this density that contains a single hydrogen atom (mass $= 1.66 \times 10^{-24}$ grams)? What is the radius of a sphere of this density that contains the mass of a single person (80 kilograms)?

72. Squeezing the Milky Way

Imagine you could squeeze all the stars in the Milky Way together, so they were just touching. For this problem, let's assume the Milky Way has 150 billion stars, each of the mass and size of the Sun. How large would the resulting ball of stars be? Express your answer in AU. *Hint: If you find yourself worrying about the details of how the stars are arranged, you are overthinking the problem. A rough answer is all that is needed.*

73. A star is born

Newly born stars condense out of the very thin gas and dust that fills the space between the stars. The density of this gas varies considerably from one part of the Milky Way to another, but a typical value is $\rho = 3 \times 10^{-20}$ kg/m^3. What volume (in cubic meters) of this material has to condense to make a star as massive as our own Sun (2×10^{30} kg)? Considering this volume as a sphere, what is its radius? Express the result in light-years.

74. A massive black hole in the center of the Milky Way

Stars in the center of the Milky Way (8 kiloparsecs from Earth) are orbiting around a central massive point source, which does not shine in visible light. The motions of these stars can be observed directly by watching their position change with time. One of these stars, termed S0-102, has now been observed to make a full elliptical orbit, with a period of 11.5 years. The orbit is not circular, but rather elliptical, with a semi-major axis that subtends an angle of 0.1 arcsec.

74.a Calculate the size of the semi-major axis of the orbit of S0-102 in kilometers.

74.b Calculate the mass of the central object. Express your result in solar masses.

74.c There·is another star also orbiting this object, called S0-16. It is on an extremely elliptical orbit, passing only 45 AU from the central object on its closest approach. This orbit is consistent with that of S0-102 (see part **a**), in the sense that using Newton's laws to understand the orbit leads one to infer the same central mass. Thus the mass of the central object must be enclosed within a radius of 45 AU. What is the lower limit on density that this implies for this object? Express your answer in grams per cubic centimeter.

74.d Near the Galactic center, there are roughly 1,000 stars per cubic parsec, a number much higher than the value in the vicinity of the · Sun. With this in mind, if you were to repeat the exercise in part **b** using a star 100 parsecs from the central black hole, what mass would you derive? Explain your reasoning.

75. Supernovae and the Galaxy

The mass of gas and stars in the Milky Way Galaxy is roughly 10^{11} times that of the Sun.

75.a Roughly 2% of the mass of gas and stars in the Milky Way consists of heavy elements (that is, higher on the periodic table than helium). These elements are made in supernova explosions, each of which releases about 1 solar mass of heavy elements into the interstellar medium. Calculate how many supernovae should have exploded in the history of the Milky Way to explain the observed quantity of heavy elements.

75.b Currently, a supernova goes off in the Milky Way roughly once every 100 years. If this rate were constant over the age of the universe, how many supernovae have exploded in the Galaxy? Comparing your answer with the estimate for the required number of supernovae you made in part **a**, argue whether the supernova rate should have been larger or smaller in the past.

76. Dark matter halos

In this problem, you will calculate the distribution of dark matter in our own Milky Way Galaxy. Remember that unlike the solar system, the mass of the Milky Way is not all concentrated to its center. The Milky Way has a flat rotation curve; that is, the rotation speed v (about 220 kilometers per second), is constant as a function of distance from the center of the Galaxy. The rotation speed at any given radius r is due to the gravitational pull of the material within the sphere, centered on the Galaxy, of radius r; the gravitational force due to the material farther out roughly averages out to zero. You may assume for this purpose that our Galaxy is spherical. In parts **a** and **b**, we are looking for an algebraic expression for an answer.

76.a Write an algebraic expression for the mass of the Milky Way contained within a radius r. *Hint: Remember that the gravity of a spherical mass acts as if all the mass were concentrated at its center.*

76.b Use your result from part **a** to derive an algebraic expression for the average density (mass per unit volume) of the Galaxy within a radius r.

76.c You will have found that the farther out one goes in the Milky Way, the larger mass one infers for it. The question then is, where is the edge of the Milky Way? The answer to this question is poorly known; the rotation curve is found to be flat as far out as it has been measured. However, other galaxies have flat rotation curves as well, implying that they also have extended halos of dark matter. We thus assume that the dark matter halos of galaxies extend out at most halfway to the next galaxy (otherwise, the halos will overlap). So now, let's calculate. The Milky Way has a rotation speed, constant with radius, of 220 km/sec. The nearest large galaxy, Andromeda, is 2 million light-years away. Use this information, and the equation for the mass you developed in part **a**, to calculate an upper limit for the total mass of the Milky Way. Express your answer in solar masses. *Hint: You may find it useful to use an approach similar to that in chapter 12 of* Welcome to the Universe, *scaling from the known speed and radius of Earth's orbit around the Sun.*

77. Orbiting galaxy

The Large Magellanic Cloud (LMC) is a small nearby galaxy in orbit around the Milky Way. It lies at a distance of 150,000 light-years from the center of the Milky Way. Its proper motion (i.e., its motion across the sky) has recently been measured to be 1×10^{-3} arcseconds per year.

77.a Calculate the orbital speed of the LMC in kilometers per second. You may assume its orbit to be circular, and make the approximation that the Sun lies at the center of the Milky Way.

77.b Calculate the time it will take the LMC to make one orbit around the Milky Way, in years. *Hint: There is a way to do this without using the results of part* **a**.

77.c Calculate the mass of the Milky Way out to the radius of the LMC from this information; express your result in solar masses. *Hint: You may scale from the result at the radius at which the Sun orbits the Milky Way.*

78. Detecting dark matter

This is a more challenging problem.

This problem asks you to design a detector of dark matter. We will work under the hypothesis that the dark matter is a form of weakly interacting massive particle (WIMP; that's really the term that's used!); that is, an as-yet undiscovered elementary particle. We will use what we know about the orbit of the Sun around the Milky Way to estimate just how numerous these WIMPs are around us, and we will estimate how often one of them may interact with the atoms in a specialized detector.

As we know, the Sun is orbiting the center of the Milky Way at about 220 kilometers per second. The dark matter is thought not to be moving in a regular circular orbit like the Sun, but moving every which way, a little like the atoms in a gas. So just like you feel a wind if you stick your head out the window in a car barreling down the highway, there should be a "wind" of dark matter particles hurtling past us at 220 kilometers per second.

Now, the fact that dark matter is *dark* tells us that it doesn't interact much at all with ordinary matter. But essentially all elementary particles that people know about or have contemplated interact somewhat, and many possible dark matter candidates that people have hypothesized would interact via the weak force (thus the name WIMP). One quantifies the interaction strength in terms of the cross-section, represented by the Greek letter σ, with units of area. The picture you should have in your mind is the particle as a little sphere of cross-sectional area σ moving through space. As it is moving, if the particle bumps into a proton or neutron in the nucleus of an atom, it will interact with it. The nucleus will then recoil, and we can detect and record the flash of light that results. The larger σ is, the more likely the particle is to collide with an atom in the detector.

With all this in mind, one way to detect dark matter particles is to build large vats of ultra-pure transparent material that is likely to interact with a WIMP and then wait for an interaction to happen.

This is a problem where rough calculations to a single significant figure are appropriate; we're making some simplifying assumptions that don't allow us to be more precise.

78.a In problem 76, you found that $M(<r)$, the mass of the Milky Way interior to a radius r from the center, is given by

$$M(<r) = \frac{v^2 r}{G}.$$

Moreover, the rotation curve of the Milky Way is flat; that is, the quantity v is independent of r (except in the central few thousand light-years, well inside the orbit of the Sun). We will use this information to calculate the density of matter in the vicinity of the Sun. We'll start algebraically: consider a thin spherical shell of radius r and thickness Δr centered on the Milky Way center. Calculate the mass in that shell (i.e., $M(<(r+\Delta r)) - M(<r)$) and its volume. The ratio of the two will give you the density at distance r.

78.b Now put in numbers for the position of the Sun in the Milky Way. The radius of the Sun's orbit is $r = 25{,}000$ light-years, and the speed of the rotation of the Milky Way is $v = 220$ kilometers per second. What is the density, in grams per cubic centimeter? One significant figure would be fine.

78.c Let us assume that half of the mean density in this shell is due to stars and gas, and the other half is dark matter is the form of WIMPs. Masses of elementary particles are often measured in electron volts (eV). An electron volt is in fact a measure of energy, 1 electron volt $= 1.6 \times 10^{-12}$ erg; the connection with mass is via Einstein's famous formula, $E = mc^2$. First, calculate, to two significant figures, the mass of a proton in electron volts. Next, assuming that dark matter is made of particles of mass 1,000 times that of the proton, calculate the number density (particles per cubic centimeter) of dark matter particles in the vicinity of the Sun.

78.d Let's switch to doing things algebraically at this point; we'll stick in numbers later. We'll use n for the number density of WIMPs you calculated in part **c**. The wind of WIMPs is sweeping past us at speed v. Consider a dark matter detector that consists of a cube of material of side L, containing M kilograms of material. The WIMPs will interact with the protons and neutrons (collectively known as nucleons), each of mass m, which make up the material. (This cube will be surrounded by photomultiplier tubes, which will detect the minute flashes of light when an interaction happens.)

In terms of these variables, calculate the number of WIMPs that move through the detector each second. You may assume that one face of the detector is perpendicular to the direction of the WIMP wind.

78.e Now let us calculate the probability that a single WIMP undergoes an interaction as it travels through the detector. Calculate the volume that its cross-section σ sweeps out as the particle travels through. The probability that it interacts with one of the nucleons is that volume times the number density of nucleons.

78.f Argue that the number of interactions in your detector per unit time is simply the product of the results in parts **d** and **e**. Give an expression for this rate in terms of n, v, M, m, and σ.

78.g Now put in numbers. We'll use the parameters of the Large Underground Xenon experiment (LUX), which is using 370 kilograms of liquid xenon placed in an underground laboratory in the Homestake

Mine in South Dakota. You've already calculated the number density of WIMPs expected and their speed. An estimate of the cross-section per nucleon is $10^{-45}\,\mathrm{cm}^{-2}$, a very small number indeed. The cross-section of an electron, for comparison, is 21 orders of magnitude higher. However, quantum-mechanical effects in the interaction between the WIMP and the nucleons increase this cross-section by a factor of A^2, where $A = 131$ is the atomic mass of xenon. Calculate the number of interactions that should be expected per year in the LUX detectors.

79. Rotating galaxies

One of the most beautiful galaxies in the sky is Messier 101. This is a face-on spiral galaxy. sometimes referred to as the Pinwheel galaxy. Take a look at figure 13.1 of *Welcome to the Universe,* and notice the small bulge and the prominent spiral arms traced by young blue stars. This galaxy is 30 million light-years away, and has a radius of roughly 40,000 light-years.

79.a Like our Milky Way, the Pinwheel galaxy rotates around its center at a speed of 200 km/sec at a radius of 20,000 light-years from the center. Thus if we watched long enough, we should see it appear to rotate in the sky. The Gaia satellite will be able to measure positions of stars to an accuracy of 10 micro-arcseconds. How long would you have to watch to see a 10 micro-arcsecond motion of stars in the Pinwheel galaxy due to rotation? Express your answer in years.

79.b Just to get a sense of how small the angle of 10 micro-arcseconds is, calculate the distance at which a dime (diameter 2 centimeters) subtends this angle. Express your answer in kilometers, and compare with the distance from Earth to the Moon.

80. Measuring the distance to a rotating galaxy

Consider a rotating spiral galaxy whose disk is inclined relative to the line of sight by $45°$. Thus the motions of its stars, moving in circles around the center, include a component in the plane of the sky (thus measurable by their change in position, as in part **a** of problem 79) and a component along the line of sight (thus measurable by Doppler shift from the spectrum). Suppose that the Doppler shift measurement of a group of stars implies a speed for them of 150 km/sec, and the motion in the plane of the sky of those same stars is 10 micro-arcseconds per year. How distant is the galaxy from us? Express your answer in megaparsecs.

14

THE EXPANSION OF THE UNIVERSE

81. The Hubble Constant

In this problem, you are going to work directly from observational data to estimate the expansion rate of the universe.

Figure 3 shows the spectra of a star and four galaxies. For each of them, we indicate the measured brightness, in units of Joules per square meter per second. Assume that each of them has the same luminosity as that of the Milky Way (10^{11} times the luminosity of the Sun, or 4×10^{37} Joules/sec). We will use the information here to determine both the distance and redshift of each galaxy, and thus infer the Hubble Constant, which quantifies the expansion rate of the universe.

81.a Determine the distance to each of the four galaxies, using the inverse square relation between brightness and luminosity. Express your answers both in meters and in megaparsecs, and give two significant figures.

81.b The spectrum of each galaxy shows a pair of strong absorption lines of calcium, which have rest wavelength $\lambda_0 = 3,935$ Ångstroms and $3,970$ Ångstroms, respectively. The wavelengths of these lines in these galaxies have been shifted to longer wavelengths (i.e., redshifted) by the expansion of the universe. As a guide, the spectrum of a star like the Sun is shown in the upper panel; the calcium lines are at zero redshift. Measure the redshift of each galaxy. That is, calculate

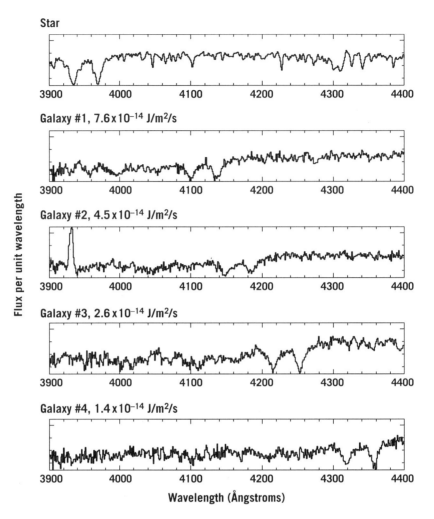

Figure 3. Figure for problem 81. Shown are the measured spectra of a star and four galaxies, all of which have strong absorption lines from Calcium. The observed brightness of each galaxy is indicated.

the fractional change in wavelength of the calcium lines. *Hint: The tricky thing here is to make sure you're identifying the right lines as calcium. In each case, they are a close pair; for galaxy 2, they are the prominent absorption dips between 4,100 and 4,200 Ångstroms. Measure the redshift by measuring both of the lines in each galaxy. (In each case, the two lines should give the same redshift, of course!) In this problem, give your final redshift to two significant figures. Do the intermediate steps of the calculation without rounding; rounding too early can result in errors.*

81.c Given the redshifts, calculate the velocity of recession for each galaxy in kilometers per second, and in each case use the distances to estimate the Hubble Constant in units of kilometers per second per megaparsec. You will not get identical results from each of the galaxies, due to measurement uncertainties (but they should all be in the same ballpark), so average the results of the four galaxies to get your final answer.

82. Which expands faster: The universe or the Atlantic Ocean?

The expansion of the universe discovered by Hubble really only takes place on physical scales larger than galaxies, but for the purposes of this exercise, imagine that the proportionality between distance and recession velocity holds on scales comparable to the size of Earth. Calculate by what distance a patch of intergalactic space as wide as the Atlantic Ocean (width 6,000 km) would grow in a year under the expansion of the universe. Compare with the actual growth of the Atlantic Ocean, due to plate tectonics (a completely different physical mechanism), which averages an inch a year. Which is larger?

83. The third dimension in astronomy

When we look at an astronomical image, it appears two dimensional; we have no depth perception and cannot immediately see how distant objects are. Astronomers use three basic methods to measure the distances to objects: parallax, the inverse square law, and redshifts. Write an essay describing what each of these methods is (using equations, diagrams, or both where appropriate); describe the types of objects for which and the range of distances over which each can be used. Describe the difficulties involved with each method.

84. Will the universe expand forever?

This is a more challenging problem.

We live in an expanding universe. However, the gravity of matter in the universe is pulling back to slow the expansion. In this problem, we will ignore dark energy and ask whether the gravity of matter acting alone would be strong enough to eventually stop, and reverse, the expansion. We will do this by considering the gravitational force on a galaxy a certain distance from us and comparing its velocity with the gravitational escape speed: is it traveling fast enough to escape? We will find a constraint on the mean density ρ (mass per unit volume) of the universe. In parts **a** and **b**, you will be doing algebra, and only plug in numbers in parts **c** and **d**.

84.a Consider a galaxy a distance r from the Milky Way. The distance r is much greater than the size of the Milky Way itself. Relative to the Milky Way, it feels the gravitational pull of all the material within a sphere of radius r centered on the Milky Way. The gravitational forces due to matter outside the radius r pull in all directions, and so have no net effect. In terms of ρ and G, calculate the mass of the material within that sphere, and determine the escape speed from that radius.

84.b The universe is expanding, so this galaxy is moving away from us at a speed $v = H_0 r$, where H_0 is the Hubble constant. The galaxy will escape (i.e., will continue traveling away forever) if that speed is greater than the gravitational escape speed. Using this idea, determine a value of the density of the universe ρ, above which the gravity is strong enough to cause the galaxies moving outward to eventually slow down, reverse motion, and fall back again. Does your result depend on the specific galaxy you chose to do the calculation?

84.c Now it is time to plug in numbers. Calculate this density (the "critical density"), using a Hubble Constant of 67 km/sec/Mpc. Express your answer in units of kilograms per cubic meter; one significant figure is fine. If this density were completely in the form of hydrogen atoms, how many hydrogen atoms would there be in the average cubic meter of space?

84.d The mass of stars of the Milky Way is roughly 10^{11} times the mass of the Sun. Galaxies are roughly 15 million light-years apart from one another. Assuming that all galaxies have a mass equal to that of the Milky Way, calculate the mean density of the universe in kilograms per cubic meter. Is the value you have found larger or smaller than that in part **c**?

85. The motion of the Local Group through space

This is a more challenging problem.

The cosmic microwave background (CMB) is observed to be uniform to an impressive degree. The principal deviation from perfect uniformity is a dipole pattern across the sky, with an amplitude of 1.23×10^{-3}. That is, in one direction, the intensity I is higher than the average by an additional fractional part of 1.23×10^{-3}. In the opposite direction, the intensity is lower by the same amount, and it varies smoothly in between. This is interpreted as a Doppler effect, caused by the motion of Earth relative to the frame defined by the material emitting the CMB itself. In this problem, you may ignore the 30 km/sec offset between the frame of reference of Earth and the Sun.

85.a How fast is Earth moving relative to the CMB? Give your answer in km/sec. The Doppler formula for intensity is exactly the same as that for wavelength.

85.b We know that Earth (and the Sun) are in orbit around the center of the Milky Way, and the Milky Way and the Andromeda Galaxy are falling toward each other. The velocity vector of the Sun relative to the center of mass of the Local Group of galaxies is 300 km/sec, in a direction approximately *opposite* to the direction of our motion relative to the CMB. Thus the motion of the Local Group of galaxies relative to the CMB is the sum of the result you found in part **a** and 300 km/sec, i.e., 660 km/sec.

Why is the Local Group moving? It is pulled by the gravity of clusters and superclusters of galaxies in our vicinity (i.e., within 50 megaparsecs or so). In particular, about half the motion is due to the nearest supercluster of galaxies, the Virgo Supercluster, centered a distance of roughly 18 Mpc from us. Although the distance between the center of the Virgo Supercluster and us will of course grow with the expansion of the universe, for this calculation, you may take it as constant. Assuming that this force (which you should approximate as constant with time) has acted for the age of the universe, and that the initial relative velocity between the Local Group and the Virgo Supercluster was zero, calculate the inferred mass of the Virgo Supercluster (out to the distance of the Local Group). Express your result in solar masses. Note that you are calculating a full gravitational mass here, dark matter included!

85.c The mass you've just calculated in part **b** must be an excess over the average background mass density of the universe, as the gravitational pull of the uniformly distributed component is zero. The galaxy density in that sphere is observed to be about a factor of 2 larger than the average. Assuming that the dark matter density is also a factor of 2 higher, you can calculate the mean mass density of the universe. Compare with the critical density, and comment.

15–16

THE EARLY UNIVERSE
AND QUASARS

86. Neutrinos in the early universe

Neutrinos are elementary particles that are involved in interactions involving the weak force. They were produced in profusion in the first seconds after the Big Bang, and these primordial neutrinos are still around today. However, neutrinos interact very rarely with ordinary matter (that's why they call it the *weak* force), and so detecting these neutrinos is a real challenge. In this problem, we will consider just how common neutrinos are in the universe. Neutrinos come in three flavors (called electron, muon, and tau neutrinos); in what follows, we are using the numbers for all three together.

86.a Neutrinos from the early universe pervade space with a number density roughly 1.5×10^9 times larger than that of ordinary atoms. The ratio of the average mass density of ordinary atoms in the universe to the critical density (Ω_{atom}) is about 4%. Given that most of the atoms in the universe are hydrogen, calculate the number density of neutrinos (i.e., in units of neutrinos per cubic centimeter).

86.b The 2002 Nobel Prize in Physics was awarded for the discovery that neutrinos have mass, and that the different flavors of neutrinos have different masses from one another. The exact values of their masses are unknown, but are thought to be less than 0.1 electron volt. Assuming that the average neutrino mass over the three flavors is indeed 0.1 electron volt, calculate the mass density of the universe in neutrinos, compare with the critical density of the universe (the value above which the universe would eventually recollapse), and comment. One electron volt is a truly tiny measure of mass, equivalent to 1.8×10^{-36} kilograms.

87. No center to the universe

Explain how it is that if the universe came from a Big Bang, there is no center to the expansion. A few short paragraphs is sufficient.

88. Luminous quasars

As described in chapter 16 of *Welcome to the Universe,* quasars are tremendously luminous; a typical quasar has a luminosity 10^{12} times that of the Sun. Quasars are powered by black holes: a typical quasar has a black hole with a Schwarzschild radius of 3×10^8 kilometers. *Hint: Calculations will be easier in this problem if you scale from your knowledge of the properties of the Sun, given in the Useful Numbers and Equations.*

88.a What is the mass of a black hole of Schwarzschild radius 3×10^8 kilometers? Express your answer in solar masses.

88.b What is the surface temperature of a spherical blackbody of radius equal to the Schwarzschild radius of the black hole in part **a**, with a luminosity 10^{12} times that of the Sun?

88.c A main sequence G star (i.e., one with the properties of the Sun) at a distance of 3,000 light-years and a quasar with luminosity 10^{12} that of the Sun, have the same apparent brightness. What is the distance to the quasar? Express your result in light-years.

89. The origin of the elements

One of the great developments of twentieth-century astrophysics has been an explanation of the origins of the atomic elements. Describe this qualitatively in a brief essay. In particular, make sure that your answer includes discussions of

- The origin of elements in the first minutes after the Big Bang. Which elements are synthesized in this process?
- The formation of elements in stars. The distinction between low-mass stars and high-mass stars in this context.
- The significance of the element iron.

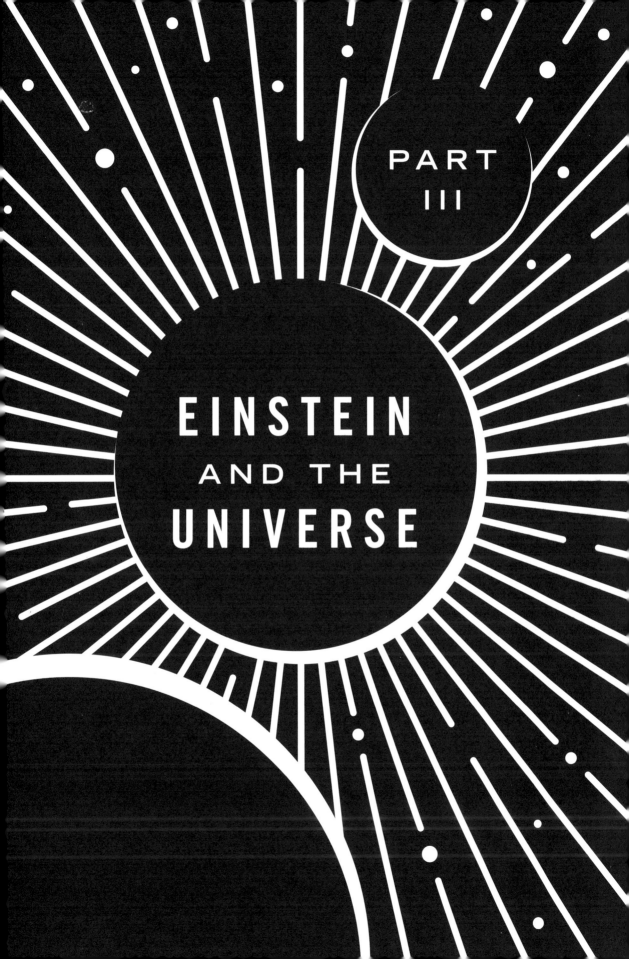

PART
III

EINSTEIN
AND THE
UNIVERSE

17–18

EINSTEIN'S ROAD TO SPECIAL RELATIVITY

90. Lorentz factor

In this problem, we will explore the nature of the function $y = \sqrt{1 - \frac{v^2}{c^2}}$. This quantity, sometimes called the "Lorentz factor," is the factor in special relativity by which an astronaut moving by at speed v ages. That is, you age n years while you observe that the astronaut ages $n \times y$ years. Note that the quantity y is always less than or equal to unity.

We start by exploring the behavior of this function for small values of v, as well as those close to the speed of light. To do this, we'll need to develop a few mathematical tools. If we define $x = v/c$, the Lorentz factor can be written $y = \sqrt{1 - x^2}$.

90.a For very small velocities, $v \ll c$ (i.e., v is *much less* than c), and we expect y to be very close to (but slightly less than) 1. Thus we write $y = 1 - \epsilon$, where $\epsilon \ll 1$. Our exercise will be to determine ϵ. Solve the equation:

$$y = 1 - \epsilon = \sqrt{1 - x^2}$$

for ϵ in terms of x. Start by squaring both sides of the equation, and then recognize that if ϵ is small, ϵ^2 is tiny, and additive terms involving ϵ^2 can be neglected. With the value of ϵ in terms of x in hand, now express the Lorentz factor y in terms of v and c, for $v \ll c$.

90.b Now let's take the opposite limit, namely, speeds very close to the speed of light. This time, we'll write $x = v/c = 1 - \alpha$, where now

$\alpha \ll 1$. Plug into the equation for the Lorentz factor; similar to part **a**, if α is small, then α^2 is really tiny and can be neglected. Thus write an expression for the Lorentz factor in this case.

90.c Now we're ready to plug in some numbers. Draw a graph of the Lorentz factor as a function of velocity, where the x-axis ranges from 0 to the speed of light c, and the y-axis ranges from 0 to 1. Plug in many values of v and calculate the value of the Lorentz factor, and plot them up. You may draw your graph by hand or use graphing software on a computer.

Next, plot on the same graph the two approximations you calculated, for small velocities and large velocities, in parts **a** and **b**. Over what range of velocities does each do a decent job of approximating the Lorentz factor? That is, over what range does each approximation give answers within 10% of the correct value?

90.d What is the value of y for $v = 100$ km/hour (a typical highway speed), $v = 30$ km/sec (the speed of Earth traveling around the Sun), $v = 0.6c$, $v = 0.8c$, and $v = 0.9999995c$? Use the approximations you've just developed, as appropriate. If you find yourself rounding off and saying that $y = 1$ or $y = 0$, you've rounded too much! In particular, the emphasis here is on the difference of your result from $y = 1$ (for low velocities) or from $y = 0$ (for high velocities).

91. Speedy muons

A muon is a particle very much like an electron, only with more mass. Unlike electrons, muons are unstable; they have a half-life of only 2.2×10^{-6} seconds, after which they decay into an electron, a muon neutrino, and an anti-electron neutrino. Very fast-moving muons are produced in Earth's upper atmosphere (i.e., 100 km above Earth's surface) when high-energy cosmic rays (produced ultimately from distant supernova explosions) collide with atoms in the atmosphere. These muons come whizzing down to be detected on the surface. A typical atmospheric muon is moving at 0.9999995 the speed of light; i.e., $(1 - 5 \times 10^{-7})c$. Taking into account the special relativistic slowing of time, how far could such a muon travel before decaying (i.e., before one half-life is over)? Now do the same calculation, not taking into account the special relativistic effects. Discuss: Does the fact that we observe muons that were produced 100 km away give any support to Einstein's prediction that time slows in a fast-moving reference frame?

92. Energetic cosmic rays

We are familiar with the equation for kinetic energy: an object of mass m moving at speed v has kinetic energy $\frac{1}{2}mv^2$. However, the equation isn't right when one gets to very high speeds. The correct version of this equation in special relativity is that the total energy of the object, the sum of its rest-mass energy (i.e., the mc^2 part) and its kinetic energy, is equal to mc^2/y, where $y = \sqrt{1 - \frac{v^2}{c^2}}$ is the Lorentz factor.

One of the most energetic cosmic rays ever observed was detected in 1991, and had an energy of about 3×10^{20} electron volts; this particular particle has been given the nickname the "Oh-My-God particle," thereby proving that astronomers have a corny sense of humor. It was probably a proton, whose rest-mass energy (mc^2) is about 10^9 electron volts. How fast was the proton moving? Express your answer as a fraction of the speed of light. This fraction will be exceedingly close to, but not exactly, unity; thus express your answer as $c(1 - \alpha)$ as in problem 90.**b**, and give the value of α.

93. The *Titanic* is moving

Here we're going to play relativistic games with the great ship *Titanic*.

93.a The *Titanic* was 882.5 feet long from bow to stern. If it sails past a dock at high speed, it will seem to people on the dock to be slightly shorter than 882.5 feet because of special relativity. How fast would the *Titanic* have to be sailing past the dock so that people on the dock would judge it to be 882.4 feet long? Why don't we notice the special relativity shortening effect in everyday life?

93.b Now start with the *Titanic* sitting at the dock and raise it to an altitude of 51.3 kilometers over New York City. Then drop it. Ignoring air resistance, when it hits the ground, it will be traveling at a velocity of 1 kilometer per second—I'm flying, Jack!

Titanic had a mass of approximately 45,000 metric tons, where 1 metric ton is 10^6 grams. Calculate the energy of the explosion when it hits in equivalent kilotons of TNT. A kiloton of TNT, when it explodes, produces an energy of 4.2×10^{12} Joules.

93.c Calculate the amount of mass that would have to be annihilated to create a similarly energetic explosion.

94. Aging astronaut

An astronaut travels from Earth to Alpha Centauri (4 light-years away) at 80% of the speed of light. How much older will the astronaut be when she arrives at Alpha Centauri? Express your answer in years.

95. Reunions

You go to your fiftieth college reunion. There you meet your college roommate who has spent the entire time since graduation traveling to a nearby star and back in a spaceship at constant speed. Your roommate says that it is nice to be at the fortieth reunion. You say, "No, it's the fiftieth reunion!" Your roommate says, "But I've kept careful track on my spaceship clock, and it's been exactly 40 years since our graduation." How fast was his spaceship going? Express your answer as a fraction of the speed of light. How far away was the star he visited (in light-years, as measured by you)? (A couple of students of ours [twins actually] once wrote a science fiction story with this plot!)

96. Traveling to another star

The star Gliese 581 has a planet in orbit around it that is in the habitable zone (i.e., water would likely be liquid on its surface). Consider an advanced civilization on this planet that travels to Earth in a spacecraft traveling at 80% of the speed of light relative to us. The distance between Gliese 581 and Earth is 20 light-years (as judged by us).

96.a How long do we here on Earth say the trip takes?

96.b How long do the Gliesians on the spacecraft say their trip lasts?

96.c How far apart do the Gliesians on the spacecraft say their planet and Earth are?

97. Clocks on Earth are slow

Earth goes around the Sun at 30 km/sec, and therefore a clock on Earth ticks slightly slower than that of an observer sitting still at the same distance from the Sun. In how many years will the two clocks differ by 1 second? Give your answer to the nearest year.

98. Antimatter!

Two 10-metric-ton trucks (each with mass $m = 10^4$ kilograms), each going 100 kilometers per hour in opposite directions, have a head-on collision. Each truck has a kinetic energy of $\frac{1}{2}mv^2$ according to Isaac Newton. Since the velocity is small relative to the speed of light, Newton's formula is quite accurate. When the trucks collide, the energy of the resulting explosion is equal to the total original kinetic energy of the two trucks. Now suppose that a 10-metric-ton truck made of matter collides with a 10-metric-ton truck made of antimatter, so that the two trucks annihilate, causing an explosion in which the total mass of the two trucks is converted into energy according to Einstein's formula $E = mc^2$. Assume that the matter and antimatter trucks hit at a speed much less than the speed of light, so their kinetic energy is negligible relative to their rest-mass energy. Calculate the ratio of the energy of collision of the matter and antimatter trucks to that of the two normal trucks in the head-on collision described above. If you see a matter and an antimatter truck about to hit each other, should you try to get far away?

99. Energy in a glass of water

Consider an 8-ounce (1/4 kilogram) glass of water. If all the hydrogen in that water underwent thermonuclear fusion to create helium, how many Joules of energy would be produced? Remember that 0.7% of the rest-mass energy of hydrogen is converted to energy in fusion.

Each of the 10^7 residents of New York City use roughly 5,000 kilowatt-hours of electrical energy per year. If we could convert all the hydrogen in water to helium and convert the resulting released energy to electricity, how many glasses would be required to power New York City each day? *Hint: A kilowatt-hour is a unit of energy. A watt is a unit of power, 1 Joule per second, and a kilowatt is 1,000 watts.*

100. Motion through spacetime

In a three-dimensional spacetime diagram of the solar system, in which time is shown as the vertical coordinate, the worldline of the Sun looks like a pole standing straight up. What does the worldline of Earth look like? Describe it qualitatively; feel free to include a sketch.

101. Can you go faster than the speed of light?

You sit in a spacecraft and shine a lightbeam toward the forward wall. Considering this thought experiment, explain how the postulates of special relativity do not allow the spacecraft to travel faster than the speed of light.

102. Short questions in special relativity

We're looking for answers of a few sentences at most for each question.

102.a Observers moving relative to one another can disagree on the simultaneity of distant events. TRUE or FALSE?

102.b What is Einstein's most famous equation? What do the different terms in the equation mean? What are some of the implications of this equation?

102.c Explain how time travel to the future is possible with present-day technology. Explain how we know this experimentally. Who has traveled the furthest into the future? How far has he time-traveled?

102.d Cowboy Bob and Cowboy Bill entered opposite ends of Main Street in Dodge City, each traveling toward the center of town at half the speed of light, as measured by the sheriff sitting still in the town center. According to the sheriff, Bob and Bill both fired their laser pistols simultaneously at each other as they entered opposite sides of the town. Cowboy Bob claims that Bill shot first, while Cowboy Bill claims that Bob shot first. Is it possible that everyone is telling the truth? Explain qualitatively in a sentence or two what principles may be involved.

102.e An astronaut sits in the exact middle of his rocket ship and shines laser beams toward the front and back. He observes that they reach the front and back simultaneously. If he is moving past you at a speed of 60% the speed of light, do you see both laser beams hitting the front and back of his rocket simultaneously? Explain, and draw a spacetime diagram to make your point.

19

EINSTEIN'S GENERAL THEORY OF RELATIVITY

103. Tin Can Land

Figure 4 shows a map of Tin Can Land. You can copy this diagram and cut it out and assemble it into its tin can shape. The top and the bottom of the can are circles, and the rectangle curls up to be the cylindrical side of the can. The triangle and square are inhabitants of Tin Can Land—they live on the surface of the can like flatlanders live on the surface of a plane. If the triangle wants to go visit the square, she can go directly along the path indicated by the straight line. That path is a geodesic: a geodesic path is one that bends neither to the left nor the right. A little car could drive along this path along the surface of the tin can without turning its steering wheel to the left or right.

103.a Draw three other geodesic paths from the triangle to the square, and label them 1, 2, 3. These are other paths a little car could take to drive along the surface of the tin can from the triangle to the square without turning the steering wheel to the left or right. It may be helpful to copy the figure and assemble it into the tin can shape to see how to do this.

103.b Which geodesic path is the shortest path from the triangle to the square?

103.c Draw on your map a triangle ABC with three right angles in it. Label each point, A, B, C, and show the geodesic paths connecting them on the map.

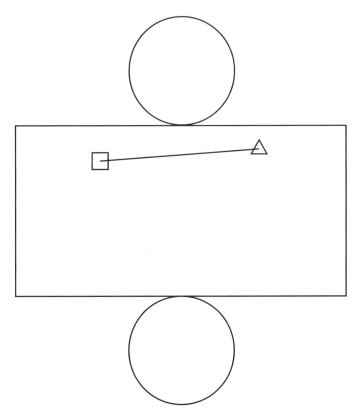

Figure 4. Figure for problem 103. The unrolled surface of Tin Can Land is shown, on which you will explore the nature of geodesics. You can copy this diagram and cut it out and assemble it into its tin can shape. The top and the bottom of the can are the circles in the figure, and the rectangle curls up to be the cylindrical side of the can.

103.d Explain how it is possible to have a right triangle with three right angles in it on the surface of the tin can, violating the tenets of Euclidean geometry, even through the above map is plotted on the plane without distortion.

104. Negative mass

If you simultaneously release a ball of 1 kilogram in mass and a ball of 2 kilograms in mass, they will both fall to the floor and hit the floor at the same time. Suppose you have a ball of *negative* 1 kilogram in mass. If you release it, will it fall and hit the floor, or will it "fall" upward and hit the ceiling? Answer this question as if you were Newton and then Einstein. What would Newton say? And why? What would Einstein say and why? State their reasoning using equations if necessary.

105. Aging in orbit

This is a more challenging problem.

In addition to the slow aging we have learned about in special relativity due to an object moving at a finite speed, there is a slow aging effect in general relativity caused by gravity itself. To understand this latter effect, consider a photon escaping from Earth's surface to infinity. It loses energy as it climbs out of Earth's gravitational well. As its energy E is related to its frequency ν by Einstein's formula $E = h\nu$, its frequency must therefore also be reduced (it is a sort of redshift), so observers at a great distance $r = \infty$ must see clocks on the surface ticking at a lower frequency as well.

Therefore, an astronaut orbiting Earth ages differently from one sitting still far from Earth for two reasons; the effect of gravity and the time slowing due to motion. In this problem, you will calculate both these effects and determine their relative importance.

105.a The escape speed from an object of mass M if you are a distance r from it is given by

$$v_{escape} = \sqrt{\frac{2GM}{r}}.$$

That is, if you are moving this fast, you will not fall back to the object, but will escape its gravitational field entirely.

Karl Schwarzschild's solution to Einstein's field equations of general relativity shows that a stationary, nonmoving clock at a radius $r > r_\oplus$ from Earth will tick at a rate that is

$$\sqrt{1 - \frac{1}{c^2}\frac{2GM_\oplus}{r}} = \sqrt{1 - \frac{v_{escape}^2}{c^2}}$$

times as fast as one located far away from Earth (i.e., at $r = \infty$). Note how much this expression looks like the equivalent expression from special relativity for aging. Here $\oplus$ is the symbol for Earth, so $r_\oplus$ is the radius of Earth, and $M_\oplus$ is its mass. Using the formula above, calculate the rate at which a stationary clock at a radius r (for $r > r_\oplus$) will tick relative to one at the surface of Earth. Is your rate greater or less than 1? If greater than 1, this means the high-altitude clock at $r > r_\oplus$ ticks faster than one on the surface; if less than 1, it means the high-altitude clock ticks slower than a clock on the surface.

105.b Now consider an astronaut orbiting at $r > r_\oplus$. What is her orbital velocity as a function of r? Because she is moving with respect to a stationary observer at radius r, special relativity says that her clock is ticking slower. Calculate the ratio of the rate her clock ticks to that of a stationary observer at radius r.

105.c Now combine the results of parts **a** and **b** to determine an expression for the ratio of the rate at which the orbiting astronaut's clock ticks to a stationary clock on the surface of Earth, as a function of the radius r at which she orbits. You may ignore the small velocity of the clock on the surface of Earth due to Earth's rotation.

105.d We need to develop some useful mathematical tricks to simplify the expression in part **c**. First, show that if $x \ll 1$, then $\sqrt{1-x} \approx 1 - \frac{1}{2}x$.
Hint: Take the square of both sides of the expression, and see what you get.

105.e Here's another similar approximation we'll need. Demonstrate that if $|x| \ll 1$, then

$$\frac{1}{1-x} \approx 1 + x.$$

Hint: Multiply both sides by $1 - x$.

Also show that if both $|x| \ll 1$ and $|y| \ll 1$, then

$$(1-x)(1-y) \approx 1 - (x+y).$$

105.f Use the approximations from parts **d** and **e** to derive a final expression, of the form $1 - \alpha$, for the rate of a ticking clock orbiting with the astronaut relative to a clock on Earth's surface. Demonstrate that your quantity α is indeed much smaller than 1.

105.g Using your result from part **f**, calculate the radius r at which the clock of the orbiting astronaut ticks at the same rate as a stationary one on the surface of Earth; express your result in Earth radii and kilometers. Will an astronaut orbiting at a smaller radius age more or less than one who stayed home? Therefore, do astronauts on the Space Shuttle (orbiting 300 km above Earth's surface) age more or less than one staying home?

106. Short questions in general relativity

We're looking for answers of a few sentences in each question, at most.

106.a Einstein used curved spacetime to explain gravity. TRUE or FALSE?

106.b What is a geodesic?

106.c Describe Einstein's thought experiment comparing a lab on Earth's surface with a lab on an accelerating rocket in interstellar space, and name the theory to which it led.

106.d What two important predictions did Einstein make in 1915 with his general theory of relativity, which differed from Newton's theory of gravity? Which prediction correctly explained an existing observation, and which one called for a new observational test? What was that observational test, when was it carried out, and what was the result?

20

BLACK HOLES

107. A black hole in the center of the Milky Way

Black holes exist! The closest big one ("supermassive," in the jargon) sits at the center of the Milky Way Galaxy, a mere 25,000 light-years from us. We will calculate its properties here. Note the related problem 74.

107.a There is a star in orbit around this black hole, with a semi-major axis of 1,000 AU. The period of this orbit is 16 years. What is the mass of the black hole? Express your answer in solar masses.

107.b The center of our Galaxy is 8,000 parsecs, or about 25,000 light-years, from Earth. What angle does the semi-major axis of the star's orbit (i.e., 1,000 AU) subtend as seen from Earth? Express your answer in arcseconds.

107.c What is the radius of the event horizon (the Schwarzschild radius) of the black hole? Express your answer in kilometers.

108. Quick questions about black holes

108.a A black hole will increase in mass and in size if material falls into it. Can a black hole shrink in mass as well? Explain.

108.b John Wheeler was one of the most creative physicists of our time. What astronomical term is he famous for coining? What ideas did he give Jacob Bekenstein and Richard Feynman that they later used in their research?

108.c Black holes emit Hawking radiation. TRUE or FALSE?

108.d Black holes are completely stable, and can only increase in mass with time. Thus they will live forever. TRUE or FALSE?

109. Big black holes

In this problem, we are going to explore the Schwarzschild radii and tidal forces associated with black holes of different masses. The Schwarzschild radius of a black hole of mass M is given by $\frac{2GM}{c^2}$.

Consider a small object of radius r a distance d from a second, much larger object of mass M. The difference between the gravitational acceleration along the line separating the two objects at the surface and the center of the first object, due to the second object, is:

$$\text{gravitational tidal acceleration} = \frac{2GMr}{d^3}.$$

(This expression is derived in problem 63, in a somewhat different context.) This difference in acceleration works to pull the object apart, and is referred to as a tidal acceleration. Calculate the Schwarzschild radius and gravitational tidal acceleration at a distance $d = 2R_{\text{Schwarzschild}}$ for an $r = 1$ meter object (about the size of a human being), for black holes of 30 solar masses (roughly the mass of each of the two black holes recently discovered colliding with each other), 4×10^6 solar masses (the mass of the black hole at the center of the Milky Way Galaxy), and 10^9 solar masses (the mass of the black holes thought to power the most luminous quasars).

110. A Hitchhiker's challenge

This is a more challenging problem.

> A full set of rules [of Brockian Ultra Cricket, as played in the higher dimensions] is so massively complicated that the only time they were all bound together in a single volume they underwent gravitational collapse and became a Black Hole.
>
> —Chapter 17 of *Life, the Universe and Everything*, the third volume of the *Hitchhiker's Guide to the Galaxy* series (1982, Douglas Adams)

A quote like that above is crying out for a calculation. In this problem, we will answer Adams's challenge and determine just how complicated these rules actually are.

An object will collapse into a black hole when its radius is equal to the radius of a black hole of the same mass; under these conditions, the escape speed at its surface is the speed of light (which is in fact the defining characteristic of a black hole!). We can rephrase the above to say that an object of a given mass will collapse into a black hole when its density is equal to the density of a black hole of the same mass.

110.a Derive an expression for the effective density of a black hole of mass M. Treat the volume of the black hole as the volume of a sphere of radius given by the Schwarzschild radius. As the mass of a black hole gets larger, does the density grow or shrink?

110.b Determine the density of the paper making up the Cricket rule book, in units of kilograms per cubic meter. Standard paper has a surface density of 75 grams per square meter, and a thickness of 0.1 millimeters.

110.c Calculate the mass (in solar masses), and radius (in AU) of the black hole with density equal to that of paper.

110.d How many pages long is the Brockian Ultra Cricket rule book? Assume the pages are standard size (8.5 inches × 11 inches). For calculational simplicity, treat the book as spherical (a common approximation in this kind of problem). What if the rule book were even longer than you have just calculated? Would it still collapse into a black hole?

111. Colliding black holes!

On September 14, 2015, the Laser Interferometer Gravitational-Wave Observatory (LIGO) measured the signal from a pair of colliding black holes. The measured signal was a gravitational wave passing by the detector. In essence, LIGO uses lasers to measure, very accurately, the distance between pairs of mirrors separated by about 4 kilometers: as a gravitational wave goes by, that distance oscillates periodically by a tiny amount.

111.a Just before two black holes collided, marking the peak of gravitational radiation, the period of the oscillations was 0.004 seconds (i.e., a frequency of 250 Hz, corresponding to middle C on the piano). When the two black holes are about to merge, their event horizons are touching each other, and they are in orbit around each other at close to the speed of light. Use this information to estimate the mass of the black holes. Assume, for simplicity, that the two black holes have the same mass. Express your answer in solar masses, to a single significant figure.

111.b Detailed modeling of the gravitational wave signal showed that the two initial black holes had masses of 36 and 29 solar masses, respectively (thus our assumption that the two black holes have equal masses was pretty good). They merged to create a single black hole of mass 62 solar masses. The difference, 3 solar masses, was converted into energy and radiated in the form of gravitational waves that were eventually detected on Earth. This energy was released in about 0.02 seconds. Calculate the luminosity of this emission in watts.

111.c Calculate the visible-light luminosity of all the galaxies in the observable universe. For this purpose, you may take the number of galaxies to be 10^{11} and assume that each galaxy contains 10^{11} stars as luminous as the Sun (see problem 69). What is the ratio of the luminosity of the merging black holes to the whole observable universe?

111.d Despite this enormous energy, the observational signal here on Earth was truly tiny. The quantity that is measured is called the "strain"; it is the fractional change in the 4-kilometer distance separating the mirrors in LIGO. The measured strain was 10^{-21}, at its peak. Calculate how large a change of distance this corresponds to; express your answer in Ångstroms. To get a sense of just how small this is, compare with the radius of a proton, about 10^{-5} Ångstroms.

112. Extracting energy from a pair of black holes

In this problem, we are going to use concepts developed by Stephen Hawking regarding the surface area of the event horizon of a black hole to determine the amount of energy that can be extracted from a pair of black holes. Hawking showed that in the collision of two black holes, the area of the final black hole's event horizon must be greater than the total initial area of the event horizons of the two original black holes.

Take two nonrotating Schwarzschild black holes at rest, each of mass M, a great distance apart. Drop them, and let them fall together under their mutual gravity until they collide to form a single black hole. Because the initial black holes are not rotating, and the two fall straight toward each other, there is no angular momentum in the system, and the final black hole will also be nonrotating.

As the two black holes fall together and collide, they will create changing gravitational fields that will produce gravitational waves traveling outward at the speed of light. These ripples in spacetime traveling outward will carry energy away from the system. The initial black holes have no electric charge, so as they fall in, no electromagnetic radiation will be emitted. They only give off gravitational radiation.

112.a The area of the event horizon of a Schwarzschild black hole is simply that of a sphere of radius given by the Schwarzschild radius. Calculate the total initial area, $A_{\text{total,initial}}$, of the two black holes' event horizons (added together) in terms of Newton's gravitational constant G, the speed of light c, and the mass of each individual black hole M.

112.b Use Hawking's result that the total area of the event horizon must increase to determine a lower limit on the mass, M_{final}, of the final black hole that is formed.

112.c Use the lower limit on the mass of the final black hole you obtained in part **b** above to set an upper limit on the mass lost by the system. Since mass-energy is conserved, and gravitational radiation is the only energy emitted by the system, the mass lost by the system must have been converted to energy. Calculate the upper limit on the amount of energy lost in the form of gravitational radiation.

112.d Divide your answer in part **c** by the total initial mass-energy of the system originally to determine an upper limit on the fraction of mass that can be converted into gravitational radiation energy in this process. Real collisions of two equal-mass nonrotating black holes will produce gravitational radiation, but less than the limit you've just calculated.

21

COSMIC STRINGS, WORMHOLES, AND TIME TRAVEL

113. Quick questions about time travel

In these two questions we are looking for short answers, of a few sentences at most.

113.a Stephen Hawking asked why, if time travel to the past is possible, are we not currently overrun by time travelers from the future. Why don't we see tourists from the future crowding historical events like the Kennedy assassination? Answer Prof. Hawking.

113.b Describe the cosmic string time machine. Describe the geometry of the space around two cosmic strings and briefly describe how this makes it possible to make a time machine.

114. Time travel tennis

This is a more challenging problem.

Here is a problem to challenge the right hemisphere of your brain. You are to draw a spacetime diagram of a time travel story involving a tennis match. It takes place in what we will call a "Ground Hog Day" spacetime, after the Bill Murray movie in which he relived the same day over and over. Take a standard full-size piece of paper to represent spacetime, with one dimension of space horizontally (from left to right) and the dimension of time vertically (with the future toward the top of the piece of paper). Tape

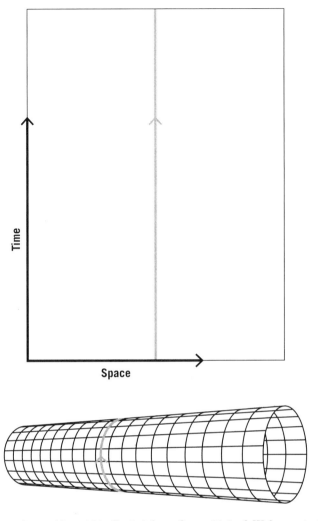

Figure 5. Figure for problem 114. Copied from figure 21.3 of *Welcome to the Universe*.

the top of the paper to the bottom of the paper to turn it into a horizontal cylinder. Consider a helical worldline around the cylinder, which always goes toward the future, yet revisits the same times over and over. See figure 5.

Here is the story you are going to illustrate. There is a tennis match. The net is stationary; it is not moving in space (left or right), but it is advancing in time so its worldline will be a vertical line. It is assembled just before the tennis match starts and is disassembled just after the match is over. Thus, draw the net as a vertical line segment, near the center of the page and about 4 inches long. You are born near the bottom left of the page. As your worldline extends to the future, you watch the tennis match. Then you go over the top of the cylinder and circle back to the past, where you arrive to watch the tennis match again. You then retrieve the ball, which you find lying on the ground. You circle back in time again to be the tennis player on the left-hand side of the net. You serve the ball, which travels over the net to the other side. Your opponent returns the ball, but it fails to go over the net and falls to the ground just to the right of the net. Thus, you win the match. After the match, you walk past the ball and move to the right side. You then circle back in time to be the opponent on the right side of the net. The ball comes across the net at you, and you rush toward the net and return it. But it fails to go over the net and ends up on the ground on the right side of the net. You thus lose the match. You then move farther to the right and circle back in time to watch the match a third time as a spectator. Then you die, and your worldline ends.

By traveling in time, you are able to be both players in a tennis match and all three spectators in the match! Draw your worldline on your cylinder. Note the events on your worldline (I am born, I watch the match, I watch again, I pick up the ball, I serve, I win, I return, I lose, I watch the match a third time.) Draw the worldline of the ball. On your worldline and the worldline of the ball, draw little arrows to indicate the direction in which you are experiencing the sequence of events. Draw the worldline of the net. Mark the directions of space and time as small axes on your cylinder. Explain how the worldline of the ball shows that it is a jinn particle, with no beginning or end. When you have finished, cut the tape, unroll the cylinder, and hand in your sheet of paper. If you place a ruler horizontally across the page at the bottom and slowly move it upward, you will see the worldlines intersect its edge in a series of points that move back and forth along the ruler playing out a movie of the tennis match and the spectators! You can see the ball move across the net and then head back, but stop short. If your diagram is not working out, start over; it may take a couple of tries to get it right.

115. Science fiction
Write a fictional story, no more than 8 pages double spaced, using scientific concepts you have learned in reading *Welcome to the Universe* and doing the problems in this book. Be creative, but be scientifically accurate. You will be graded on the correctness of the scientific concepts you use, the extent to which the science is integral to your story, and your creativity and

originality. For inspiration, look at the excellent volume by Alan Lightman, *Einstein's Dreams* (Pantheon, 1993), consisting of a series of vignettes, each fancifully illustrating a concept drawn from Einstein's own science. Have fun!

We have put this assignment in the time travel chapter, as many of our students have used time travel as a basis for their stories. But you may base your story on any aspect of astrophysics.

22

THE SHAPE OF THE UNIVERSE AND THE BIG BANG

116. Mapping the universe

Rank the following objects by distance from the center of Earth, starting with the **closest** to Earth.

- Kuiper Belt Objects
- The Hubble Space Telescope
- The highest redshift galaxy known
- Earth's Moon
- The star Alpha Centauri
- The WMAP satellite
- The Cosmic Microwave Background
- The Sloan Great Wall of galaxies
- Jupiter
- The center of the Milky Way
- M31, the Andromeda Galaxy
- The star Sirius

117. Gnomonic projections

The surface of a sphere may be mapped onto a plane using the *gnomonic projection*. This works by putting a light bulb in the center of the sphere and projecting the sphere onto a plane.

On the surface of a sphere, a geodesic connecting two points (i.e., the straightest line one can draw on the sphere) is a great circle connecting the two points. If we cut the sphere with a plane passing through the center of the sphere, the intersection of that plane with the surface of the sphere is a great circle. Thus, for example, the equator is a great circle, as are all the meridians of longitude.

117.a The gnomonic projection has the following property: great circles on the sphere are mapped onto straight lines on the map. Explain why this is true, remembering some facts from your high school solid geometry.

117.b We may use the gnomonic projection to make a flat map of Earth on a plane. By the results of part **a**, any two points on the map may be connected by a straight line, corresponding to the great circle on the globe that connects those two points. The equator is a straight line, as are meridians of constant longitude. The gnomonic projection used in this way cannot map the whole sphere, but can only map at most a single hemisphere (or, in practice, slightly less). Why is this?

117.c We may use this gnomonic projection to make a flat star map of the Celestial Sphere. If we want to map the entire Celestial Sphere, we can put the sphere into a cube, and with a projection light at the center, project the sphere onto each of the cube's six faces. Each star casts a shadow onto one of the faces of the cube. In this way, we can map the Celestial Sphere onto the surface of the cube. We can then unfold the cube to make a flat map of the entire sky in six star charts: see figure 6. The top chart (the top of the cube) shows the north circumpolar stars. The bottom chart (the bottom of the cube) shows the south circumpolar stars. The four charts in the middle show, from left to right, the stars that are visible during the northern hemisphere fall, summer, spring, and winter.

The Celestial Equator is a straight line going from left to right straight across the middle of the four charts representing the four sides of the cube. In doing the problem, it may be useful to make a photocopy of this figure, cut it out, and fold it up to form a cube. You may fold it so that the stars are on the inside of the cube (as we see them), or you can fold it so that the stars are on the outside of the cube.

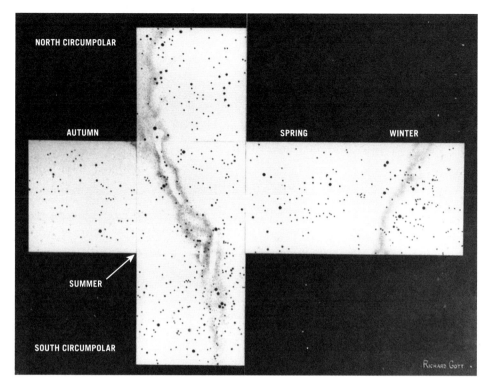

Figure 6. The night sky in gnomonic projection onto a cube. Each panel represents one face of the cube.

On the original large set of six star charts (i.e., the unfolded cube), circle and identify the following:

- The Big Dipper
- The constellation Cassiopeia
- The constellation Lyra
- The constellation Leo
- The star Sirius
- The star Alpha Centauri

To do so, you should use a conventional star chart, as may easily be found on the web, to determine where each of these constellations and stars lie.

117.d Use a dashed line to draw the ecliptic on the diagram, and label it. The ecliptic is the great circle the Sun follows relative to the background stars in its yearly journey around the Celestial Sphere. Remember that the ecliptic is tipped at an angle of 23.5° relative to the Celestial Equator. The Sun crosses the Celestial Equator at the Vernal and Autumnal Equinoxes (i.e., in the middle of the fall and spring charts).

117.e The path of the ecliptic in the star charts passes through the constellations of the zodiac. Consider how the gnomonic projection shows it. In Euclidean geometry, a point off a straight line has exactly one straight line that passes through it, parallel to the first line. That is, parallel straight lines (geodesics in plane geometry) never intersect. On a sphere, two great circles (i.e., geodesics in spherical geometry) *always* intersect. And yet the gnomonic projection maps great circles into straight lines. Comparing the ecliptic you've just drawn with the Celestial Equator, reconcile these seemingly contradictory statements.

117.f On the Celestial Sphere, a great circle route connects the star Alpha Centauri with the North Celestial Pole (where the star Polaris, the North Star, is located). This, of course, is the geodesic connecting these two points (i.e., the shortest distance on the Celestial Sphere connecting Alpha Centauri and Polaris). Draw this great circle as a solid line on the map and label it.

117.g Use a dotted line to draw the shortest route on the surface of the cube connecting Alpha Centauri and Polaris, and label the line. That is, if you were an ant walking on the surface of the cube, what shortest path would you take to get from Alpha Centauri to Polaris? Is this the same line you drew for part **f**? Why or why not?

117.h On the Celestial Sphere, we can draw a geodesic triangle (i.e., one whose three sides are each great circles) that has three right angles. It connects the South Celestial Pole with two points on the Celestial Equator that are 90° apart on the equator. Draw (with a dot-dash line) such a triangle on the star maps, and label the vertices A, B, and C.

Now we have a plane drawing of this triangle with three right angles in it, a total of 270 degrees. On a plane, the sum of the angles in a triangle is 180 degrees. And the gnomonic projection maps great circles into straight lines. So does the map of the triangle with three right angles on the sphere show a total of more than 180 degrees on the flat map? Explain.

118. Doctor Who in Flatland

The famous Doctor Who travels in the Tardis, which looks like a phone booth as seen from the outside. Open the door and go in: you find a big room on the inside that is larger in volume than its outside dimensions would indicate. Everyone says when they enter, "it's bigger on the inside!" In this problem, you will explore how this might occur if Doctor Who was a Flatlander living in Flatland—a spacetime with two spatial dimensions.

The Tardis could be as shown in figure 7: a 1 meter by 1 meter by 1 meter cubical box without a top face, taped to a 1 meter by 1 meter square hole in the flat plane of Flatland. A Flatlander entering the Tardis would slide down one of the four sides of the box to the bottom of the box. The whole thing looks like an infinite flat kitchen counter with a square kitchen sink embedded in it. It is a single two-dimensional surface. The Flatlander looks at the outside of the Tardis and thinks that since it is a square with dimensions of 1 meter by 1 meter, its area inside must be 1 square meter. But actually, the inside of the Tardis has an area of 5 square meters: each of the four sides of the sink is 1 meter square, as is the bottom of the sink. So it is bigger inside!

The Tardis is produced by the gravity of eight point masses: four at the vertices at the top corners of the sink and four at the vertices at the lower corners of the sink where two sides meet the bottom. In Flatland, mass density has units of mass per length^2. General relativity sets this mass density equal to the curvature, which has units of 1/length^2, so in Flatland, mass must be dimensionless.

The spacetime around a point mass in Flatland has the geometry of a cone; the angle deficit in radians (i.e., the difference from 2π radians, or 360°) is equal to the mass. For example, in an equilateral tetrahedron, there are three 60° triangle faces that meet at each point, a total of 180°, rather

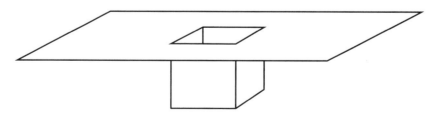

Figure 7. Figure for problem 118. This represents the geometry of a Tardis in Doctor Who (in Flatland), which appears to have a significantly larger area inside than out.

than the 360° we would expect for a point on a flat plane, giving a deficit of 360° − 180° = 180°, or π radians. So each vertex of the tetrahedron represents a point of mass π. Note that in Flatland, both positive and negative masses can exist.

118.a In units of radians, how much mass does each of the point masses in the Tardis contain?

118.b What is the total mass associated with the Tardis?

118.c Does the answer to part **b** make sense, given the shape of the geometry exterior to the Tardis?

119. Quick questions about the shape of the universe

In these questions, we are looking for short answers of a few sentences at most.

119.a The goal of Kaluza-Klein theory was to unite gravity and electromagnetism in a single theoretical framework, using four dimensions of space and one of time. TRUE or FALSE?

119.b What prediction did Gamow and his colleagues Herman and Alpher make in 1948 that was later verified by Penzias and Wilson in 1965? Why was this important?

119.c In 1922, Alexander Friedmann inferred three possible geometries for the universe. Describe them here.

119.d In a closed Friedmann universe (a model having no cosmological constant) that eventually collapses to a Big Crunch, is the sum of the angles in a triangle less than, equal to, or greater than 180°?

119.e What did Einstein think was his biggest blunder?

23

INFLATION AND RECENT DEVELOPMENTS IN COSMOLOGY

120. The earliest possible time

In this problem, we ask just how far back in time our laws of physics will allow us to push. As we've seen, gravity dominates our understanding of the very large and the very massive, while quantum mechanics describes the world of the very small. In the very early universe, the universe was exceedingly dense, and thus very small things can be very massive, and the effects of general relativity and quantum mechanics are simultaneously important. In this problem, we'll determine at what time in the history of the universe this was true. Physicists have been unsuccessful in developing a quantum theory of gravity, so this represents a limit to our current ability to ask questions about the early universe.

120.a To start, we need to learn a little quantum mechanics. You already know the equation $E = h\nu$, relating the energy E of a photon to its frequency ν, where the quantity $h \approx 2/3 \times 10^{-26}\,\mathrm{erg}\cdot\mathrm{sec}$ is Planck's constant. The Compton wavelength of a particle of mass M is the wavelength of a photon of the same energy, using the relation $E = Mc^2$. What we're really doing here is invoking the so-called wave-particle duality, which is at the heart of quantum mechanics, whereby a particle can also be thought as having wavelike properties.

Use this to derive an expression for the Compton wavelength of a particle of mass M. *Hint: To do so, you need to convert from frequency to wavelength.*

120.b So the Compton wavelength expresses the quantum nature of the particle, while the Schwarzschild radius expresses its gravitational nature. Equate the two lengths to each other to ask: what is the mass of a black hole that is equally dominated by gravity and quantum mechanics? Plug in numbers, and express this mass both in grams and in GeV (billions of electron volts).

120.c Given the mass you have just derived, you can calculate a corresponding length (i.e., the Schwarzschild radius of that mass) and a time (the time for light to travel that length). Derive expressions for each of these quantities and plug in numbers, expressing length in centimeters and time in seconds. The time you have just calculated represents the earliest time that our current laws of physics allow us to talk about the nature of the universe. At earlier times yet, both quantum mechanics and general relativity are dominant, and as we have never successfully married these two theories, we can't go further.

121. The worst approximation in all of physics

In problem 120, you calculated what are called the Planck mass and Planck length, scales beyond which one requires a quantum theory of gravity to explore. The nature of dark energy is not at all understood, but one educated guess states that it is somehow related to the nature of quantum gravity, and thus should have a density equal to the Planck density.

121.a Calculate the Planck density given the Planck mass and Planck length; give both an algebraic expression and a numerical value.

121.b Does this density estimate do a good job of explaining where dark energy comes from? That is, is this educated guess a good one? To determine this, calculate the density associated with the observed amount of dark energy. The ratio of the dark energy density to the critical density of the universe, denoted by Ω_Λ, is measured to be about 0.7. Take the ratio of the Planck density to the observed density of dark energy, and comment. If the dark energy really had a density equal to the Planck value, what would the expansion history of the universe be like? Could galaxies, stars, planets, and people have formed?

122. Not a blunder after all?

Observations of distant supernovae suggest that the expansion of the universe is accelerating today. If Einstein were alive today to hear this news, why might he say that what he thought of as his biggest blunder was not such a mistake after all? Why might Alan Guth (who is known for his theory of inflation) be happy about the supernova result as well?

123. The Big Bang

The Big Bang theory gives a natural explanation for a variety of observational facts about the universe. Write an essay describing what these facts are, and how the Big Bang model explains them. In particular, be sure to describe the importance of the redshifts of galaxies, the abundance of hydrogen and helium, different measures of the age of the universe, and all aspects of the cosmic microwave background (CMB), including its spectrum, its temperature, and its isotropy.

24

OUR FUTURE IN THE UNIVERSE

124. Getting to Mars

Suppose you want to colonize Mars. At some point, you will have to send a human flight crew to land on Mars. The most energy-efficient way to do this is to send the astronauts on a path called a Hohmann ellipse transfer orbit. For purposes of this problem, consider both Earth and Mars to have circular coplanar orbits (which is a pretty good approximation). The radii of the Earth and Mars orbits are 1 AU and 1.524 AU, respectively. The Hohmann elliptical orbit in this case is a simple Keplerian elliptical orbit with a minimum distance from the Sun of 1 AU and a maximum distance of 1.524 AU. At its minimum distance from the Sun, the Hohmann orbit intersects Earth's orbit, and at its maximum separation, it intersects Mars' orbit. So you can use the Hohmann elliptical orbit to go from Earth's orbit to Mars' orbit. To get on this elliptical orbit, you have to escape from Earth and give your rocket a kick in the direction of Earth's orbital motion. Having this increased orbital speed, your rocket will then coast outward on the Hohmann elliptical orbit until it reaches its maximum separation from the Sun of 1.524 AU. You just have to time it so that Mars will be there when you arrive! Your spacecraft will then enter Mars' atmosphere, slowing down while protected by its heat shield. You then pull out a big parachute and further slow down in the atmosphere, finally firing retro-rockets to slow you gently to a soft landing on the surface. Applying Kepler's laws, calculate how long it will take your rocket to travel from Earth to Mars. A calculator would be appropriate here. *Hint: The semi-major axis is the average of the distances between the Sun and the rocket at its closest and farthest points.*

125. Interstellar travel: Solar sails

In this problem, we will explore a futuristic way to travel between the stars.

A solar sail uses not wind (there is no such thing in the vacuum of outer space), but the pressure of light itself, to propel itself. A photon not only has energy but also momentum, equal to its energy divided by the speed of light.

When lots of photons hit a reflective surface, they bounce back, transferring twice their original momentum (do you see where the factor of 2 comes from?), to the surface itself. The total force on the surface (the rate of momentum transferred per unit time), is therefore

$$\text{Force} = 2 \times \frac{\text{Energy received by the surface per unit time}}{\text{Speed of light}}.$$

Consider a solar sail, a huge mirror, of area A a distance d from a star of luminosity L. We will consider the force on the sail from the radiation pressure of the star light hitting it.

125.a The acceleration of the sail is given by the force acting on it divided by its mass. Calculate this acceleration, assuming that the sail is made of material of mass density ρ and thickness l, and is oriented perpendicular to the line to the star (i.e., it is facing the star). Does the acceleration depend on the size of the sail?

125.b Note that we have assumed that all the mass is in the sail itself. But presumably we would like to have a payload as well. We'll assume a payload of 10^5 kilograms. We want the sail to be as large as possible, to maximize the force, but we don't want it to dominate the costs of the spaceship; we thus will require that the sail and the payload be of the same mass. Let us take a circular sail. What is the radius of the sail, if it is made of material with a density $\rho = 3\,\text{g/cm}^3$ (comparable to aluminum) and a thickness of $l = 0.1$ millimeters?

125.c Let us place this craft 1 AU from the Sun. What is the acceleration that the payload would experience from radiation pressure alone? Assuming that this acceleration stays constant, how long would it take to get to a speed of 30 km/sec per second? (In fact, as the spacecraft moves away from the Sun, the radiation pressure and thus the acceleration decrease. Incorporating these effects would make the problem much more complicated.)

126. Copernican arguments

Chapter 24 of *Welcome to the Universe* describes an argument based on the Copernican Principle to make predictions for the future longevity of various things, based on their current ages and assuming that the present moment is not special in their full lifespan.

126.a You read in the paper that the CEO of a certain company has been in charge for 39 weeks. Knowing nothing else, what would the Copernican Principle say are the 95% confidence upper and lower limits for the future longevity of the CEO's tenure? Express your answers in weeks and years, as appropriate.

126.b If we were to receive radio signals from an extraterrestrial civilization that says it is 39,000 years old, the Copernican Principle (that our time of observation is not special) would suggest there would be a 95% chance that its future longevity will be at least ____ years but less than _____ years.

126.c Explain how the Copernican Principle suggests that it is likely that you are born in a century with a population above that of a century with the median population.

127. Copernicus in action

The book, *Welcome to the Universe*, invokes the Copernican Principle (whereby we are not the center of the universe) in a variety of contexts. Answer the questions for each of the examples of the Copernican Principle given below.

127.a Earth is not at the center of the solar system. What were the advantages of the Copernican model, at the time it was introduced, over the Earth-centered model of Ptolemy? What direct observational proof do we have that Earth goes around the Sun, rather than the other way around? Did Copernicus himself have such a direct proof that his model was correct?

127.b The solar system is not at the center of the Milky Way. Why did people believe that the solar system was at the center of the Milky Way at the beginning of the twentieth century? How were they misled? What were the crucial observations that proved them wrong?

127.c The Milky Way Galaxy is not at the center of the expanding universe. In what sense does it appear that we are indeed at the center of the expansion? How would you explain to a skeptic that this is not the case?

127.d We often believe that we are living in a special time. Explain Gott's Copernican argument that the realization that the present time is not special is a powerful tool for making predictions about the future.

128. Quick questions for our future in the universe

We're looking for answers of a few sentences for each question, at most.

128.a Einstein succeeded in developing a theory that unifies the concepts of general relativity and quantum mechanics. TRUE or FALSE?

128.b The Copernican Principle suggests that the human spaceflight program is likely to last forever. TRUE or FALSE?

128.c Which occurs later in the history of our universe, the death of all the stars, or the evaporation of supermassive black holes?

128.d Explain why there is a smallest measurable time—the Planck time (5×10^{-44} seconds). You need not explain its value, just the physical reasoning that such a shortest time exists.

128.e If you had a time machine, what time in the past or future would you like to visit? Note: There are no wrong answers to this one, as long as you state a year and explain why!

129. Directed panspermia

To save the human race from extinction here on Earth by weapons of mass destruction, you decide to colonize the Milky Way Galaxy. You could design an interstellar probe containing frozen egg and sperm cells from humans and all the animal species you would need, as well as beneficial bacteria and a seed and spore bank for plants. The probe would also include instructional material for the new humans on the planet, including all relevant cultural and scientific knowledge they will need on their new planet. The probe would be launched to distant habitable planets; once it arrived, robots on board would gestate its animals and people over time and plant its seeds. Using raw material from the planet, the planets and animals would grow, just as they do here on Earth. The new inhabitants would develop a biosphere (if that didn't already exist on the planet) and a civilization in 10,000 years.

They then will be motivated to launch probes of their own, to continue the exploration of the Milky Way.

129.a A human egg cell has a diameter of 0.12 millimeters and is roughly spherical. With a density of 1 g/cm^3, what is its mass? By what factor is this less than the weight of a fully grown human being? Which is easier to send on a long space journey?

129.b Now we need to get to these distant worlds. We would need to accelerate the spacecraft up to high speed and then decelerate when approaching the planet. Project Orion and Project Daedalus are two concepts for using nuclear fusion to achieve spacecraft velocities of, say, 1% the speed of light. Daedalus planned to ignite pellets of deuterium and helium-3 using electron beams in a reaction chamber, while Orion imagined exploding hydrogen bombs at the backside of a spacecraft. In either case, the push from these explosions propels the spacecraft forward. While the technology to actually make this work is not yet in hand, these options do appear feasible.

We have estimated in *Welcome to the Universe* that there are roughly six habitable planets within 40 light-years of the Sun. How long would it take our spacecraft, moving at 1% the speed of light, to travel 40 light-years?

129.c We estimated in *Welcome to the Universe* that there are 1.8 billion habitable planets in the Milky Way. If we planted six such colonies, and after 10,000 years they each planted six colonies of their own, how long would it take to colonize all habitable planets in the Milky Way?

129.d The physicist Enrico Fermi, in thinking about this problem, famously asked, "Where are they?" If there are a number of intelligent civilizations in the Galaxy, the Copernican Principle tells us that we humans are not likely to be the first one and therefore not the most advanced. If these more advanced civilizations have also been inspired to explore the Milky Way, why have we not yet been visited, or colonized, by extraterrestrials? The Copernican Principle has an answer to this paradox. What is it?

129.e Let us remind ourselves just how difficult interstellar travel is. No human has traveled farther from Earth than the Moon. What is the ratio of a trip of 40 light-years to the distance to the Moon?

129.f The fastest humans have flown is 25,000 miles per hour. By what factor would we have to increase our speed to get to 1% the speed of light?

129.g If we wanted to build the Orion interstellar craft (one of the nuclear-powered spacecraft concepts mentioned in part **b**), we would probably need to assemble it in low-Earth orbit. Its design mass is 400,000 tons. So our challenge is to get the material to build this into low-Earth orbit.

    The Saturn V rocket is the biggest we have built thus far. It can carry a payload of 120 tons into low-Earth orbit. How many Saturn V launches would it take to assemble the Orion craft?

129.h How long have humans been traveling in space so far? By what additional factor larger would the future longevity of the space program have to be to complete the colonization of the Galaxy you calculated in part **c**?

USEFUL NUMBERS AND EQUATIONS

In what follows, we give both the modern precise values for constants (typically to four significant figures) and approximations useful for making rough calculations to "astronomical accuracy," which are often appropriate for the problems in this book.

Times and the Solar System

- 1 solar day = 86,400 seconds $\approx 1 \times 10^5$ seconds

- 1 sidereal day = 86,264 seconds

- 1 sidereal month = 27.32 solar days $\approx 2.4 \times 10^6$ seconds

- 1 year = 365.24 solar days $\approx 3 \times 10^7$ seconds

- Radius of Earth = 6.371×10^6 meters

- Radius of Moon = 1.737×10^6 meters

- Radius of Sun = 6.957×10^8 meters ≈ 109 Earth radii

- Surface temperature of Sun $\approx 5,800$ Kelvin

- Radius of Moon's orbit around Earth = $384,400$ kilometers $\approx 4 \times 10^8$ meters

- Luminosity of Sun $\approx 3.8 \times 10^{26} \frac{\text{Joules}}{\text{sec}}$

Physical Constants

- Newton's gravitational constant $G = 6.674 \times 10^{-11} \frac{\text{m}^3}{\text{sec}^2\,\text{kg}} \approx \frac{2}{3} \times 10^{-10} \frac{\text{m}^3}{\text{sec}^2\,\text{kg}}$

- Planck's constant $h = 6.626 \times 10^{-34}$ Joule sec $\approx \frac{2}{3} \times 10^{-33}$ Joule sec

- Speed of light $c = 299,792 \frac{\text{km}}{\text{sec}} \approx 3 \times 10^8 \frac{\text{m}}{\text{sec}}$

- Stefan-Boltzmann constant $\sigma \approx 5.67 \times 10^{-8} \frac{\text{watts}}{\text{m}^2\,\text{K}^4}$

- Boltzmann constant $k = 1.380 \times 10^{-23}$ Joules/K $\approx \frac{1}{7} \times 10^{-22}$ Joules/K (Yes, Boltzmann had many things named after him!)

- Mass of proton = 1.6726×10^{-27} kilogram

- Mass of neutron = 1.6749×10^{-27} kilogram

- Mass of hydrogen atom = 1.6737×10^{-27} kilogram

- Avogadro's number (the number of atoms in 1 gram of hydrogen) = 6.022×10^{23} (Note that the inverse of this number is the mass of a hydrogen atom in grams.)

- For most problems in this book, you may approximate the mass of the proton, neutron, and hydrogen atom to be the same, $\approx \frac{1}{6} \times 10^{-26}$ kilogram, or about 10^9 electron volts (1 GeV).

- Mass of electron = 9.1094×10^{-31} kilograms = 511 kilo-electron volts

Units

Michael's high-school physics teacher used to admonish his students, "Units are people too: treat them with respect!" It is vitally important in doing the problems in this book to keep track of the units in your calculations. Remember that units can be multiplied and divided just like numbers can. You should confirm that the units of your calculation make sense (thus, for example, if you are determining a distance, if the units of your answer are not length, you have done something wrong). Your units must also be consistent: if they are not, you need to convert them to make them consistent.

We mostly use the so-called MKS system of units, based on meters (abbreviated m), kilograms (kg), and seconds (sec or s), but astronomers are fond of changing units as it serves their purposes. Many problems in this book use cgs units: centimeters (cm), grams (g), and second. We indicate common abbreviations for units in what follows.

Distances are also measured in

- Ångstroms; $1\text{Å} = 10^{-10}$ meter

- microns: $1\mu\text{m} = 10^{-6}$ meter

- kilometers: $1\,\text{km} = 10^3$ meter

- Astronomical Units: $1\ \text{AU} = 1.496 \times 10^{11}$ meters $\approx 1.5 \times 10^8$ kilometers

- Light-years: $1\ \text{ly} = 9.46 \times 10^{12}$ kilometers $\approx 1 \times 10^{16}$ meters $\approx 60,000$ Astrononical Units

- parsecs: $1\ \text{pc} \approx 3.26$ light-years $\approx 200,000$ Astrononical Units

- Megaparsecs: $1\ \text{Mpc} = 10^6$ parsecs

Masses are measured in units of

- Earth masses: $1\ M_{\text{Earth}} \approx 5.97 \times 10^{24}$ kilograms

- Jupiter masses: $1\ M_{\text{Jupiter}} \approx 1.9 \times 10^{27}$ kilograms $\approx 10^{-3}$ solar masses

- Solar masses: $1\ M_{\text{Sun}} \approx 1.99 \times 10^{30}$ kilograms $\approx 333,000$ Earth masses. We also often use the symbol $\odot$ to refer to the Sun, thus a solar mass is written $1\,M_{\odot}$.

- The mass of Earth's moon is 7.348×10^{22} kilograms.

Densities are measured in kilograms per cubic meter, or grams per cubic centimeter. Liquid water at room temperature and pressure has a density of 1 gram per cubic centimeter, equal to 1,000 kilograms per cubic meter.

- In the MKS system, energy is measured in Joules;

$$1\,\text{Joule} = 1\,\frac{\text{kilogram}\,\text{meter}^2}{\text{second}^2}.$$

The cgs equivalent is

$$1\,\text{erg} = 1\,\frac{\text{gram}\,\text{centimeter}^2}{\text{second}^2} = 10^{-7}\,\text{Joules}.$$

One electron volt (abbreviated eV), a unit often used for atomic processes, is equal to 1.6×10^{-19} Joules. We often refer to multiples of electron volts, such as MeV (mega-electron volts, 10^6 eV) and GeV (giga-electron volts, 10^9 eV).

- Luminosity (energy per unit time) is measured in Joules per second (also known as watts), or in solar luminosities, $L_{\text{Sun}} = 3.8 \times 10^{26}$ watts

- Brightness is measured in Joules per second per meter^2

- Frequency (e.g., of a light wave) is measured in hertz (Hz), cycles per second

Mathematics

- $\pi = 3.1415926\ldots$; in astronomical problems to a single significant figure, it is often appropriate to approximate it as 3.

- There are 360 degrees, or 2π radians, in a circle. Degrees of angle are written with the $^\circ$ symbol.

- 1 radian $= \frac{180^\circ}{\pi} = 206,265$ arcseconds $\approx 200,000$ arcsec $\approx 57.3^\circ$.

- The small-angle formula: The angle θ in radians subtended by an object is given by its diameter s, divided by its distance d: $\theta = \frac{s}{d}$. This formula is valid for small angles, where the size of the object is much smaller than its distance. This approximation is valid for most problems in this book.

- The circumference of a circle of radius r is $2\pi r$; its area is πr^2.

- The surface area of a sphere of radius r is $4\pi r^2$; its volume is $\frac{4}{3}\pi r^3$.

Orbits and Newtonian Mechanics

- Newton's second law relates the total (net) force F on a body of mass m to the acceleration a that that force induces, $F = ma$.

- The kinetic energy of a body of mass m moving at speed v is $\frac{1}{2}mv^2$.

- The acceleration a required to keep an object moving in a circle of radius r at uniform speed v is $a = \frac{v^2}{r}$.

- The amplitude of the gravitational force between two objects of mass M and m separated by a distance r is $\frac{GMm}{r^2}$, where G is Newton's gravitational constant.

- The potential energy associated with a small body of mass m a distance r from the center of a large spherical body of mass M is

$$\text{Potential energy} = -\frac{GMm}{r},$$

where G is Newton's gravitational constant.

- Kepler's third law states that for orbits around the Sun, the period P squared is proportional to the semi-major axis of the orbit a cubed: $P^2 \propto a^3$. If period is measured in years, the semi-major axis in AU, the constant of proportionality is 1. In physical units, Newton's form of Kepler's third law is $P^2 = \frac{4\pi^2 a^3}{GM}$. When using units of years, AU, and solar masses, this becomes $P^2 = \frac{a^3}{M}$; that is to say, the constant $\frac{4\pi^2}{G} = 1$ in these units.

- The closest and farthest approach of a planet from its parent star in an orbit with semi-major axis a and eccentricity e are $a(1-e)$ and $a(1+e)$, respectively.

- The speed of an object in a circular orbit of radius r around an object of mass M is $\sqrt{GM/r}$. The escape speed from that radius is $\sqrt{2GM/r}$. The orbital speed of

Earth around the Sun ($r = 1$ AU, $M = 1$ solar mass) is 30 kilometers per second, and the escape speed from the surface of Earth (r equal to the radius of Earth, M equal to 1 Earth mass) is roughly 11 kilometers per second.

Temperature, Light, and Energy

- Temperature is measured in Kelvins: the temperature in Kelvins is that in Centigrade plus 273.15. Temperature is also measured in electron volts: the mean kinetic energy of a particle with temperature T is $\frac{3}{2}kT$, where k is the Boltzmann constant. A temperature of 10,000 Kelvin corresponds to an energy of 0.86 electron volts.

- The luminosity per unit wavelength λ per unit area of a blackbody of temperature T is given by

$$I_\lambda(T) = \frac{2\,hc^2}{\lambda^5} \frac{1}{e^{\frac{hc}{\lambda kT}} - 1}.$$

  Here h is Planck's constant, k is Boltzmann's constant, and c is the speed of light. The units of this quantity are Joules per second per meter^3. At wavelengths large compared to the peak wavelength, this simplifies to

$$I_\lambda(T) = \frac{2ckT}{\lambda^4}.$$

- The Stefan-Boltzmann law: the energy per unit time emitted by a blackbody of surface area A and temperature T is equal to $\sigma A T^4$, where σ is the Stefan-Boltzmann constant.

- The blackbody spectrum of an object of temperature T peaks at a wavelength $\lambda \approx \frac{2.9}{T}$ millimeters, if T is measured in Kelvins.

- The wavelength λ and frequency ν of a photon are related as $\lambda\nu = c$, where c is the speed of light. The energy of a photon is its frequency ν times the Planck constant h.

- Light of wavelength λ_0 emitted by an object moving toward or away from us at speed v has its wavelength shifted to λ such that $(\lambda - \lambda_0)/\lambda_0 = \pm v/c$ (the Doppler shift); the positive sign is for an object moving away, and the negative sign for an object coming toward us. This formula is valid for speeds much less than the speed of light (the nonrelativistic limit).

- The brightness b of a distant object is proportional to its luminosity L times the inverse square of its distance r to us:

$$b = \frac{L}{4\pi r^2}.$$

Astronomical Formulas and Numbers

- The equilibrium temperature of a planet of albedo A with no greenhouse effect a distance d away from a star of surface temperature T_{star} and radius r_{star} is

$$T_{\text{planet}} = (1 - A)^{1/4} T_{\text{star}} \left(\frac{r_{\text{star}}}{2d} \right)^{1/2}.$$

For a star with the radius and surface temperature of the Sun, and a planet with $A = 0$ at a distance $d = 1$ AU from the star, this gives a temperature of about 300 K.

- The parallax of a star in arcseconds (the half-angle of its total motion in the sky) due to the orbit of Earth around the Sun is the inverse of its distance in parsecs. More generally, parallax can refer to the apparent change of the position of any object due to the motion of the viewer.

- The Sun is 25,000 light-years from the center of the Milky Way, and makes a full orbit once every 2.5×10^8 years. The mass of the Milky Way within the Sun's orbit (the "solar circle") is roughly 10^{11} solar masses.

- The Hubble Law: The recession velocity v of a galaxy is equal to its distance d times the Hubble constant H_0: $v = H_0 d$.

- The Hubble constant $H_0 \approx 67$ kilometers per second per megaparsec.

- The age of the universe since the Big Bang is roughly the inverse of the Hubble constant. This gives roughly 14 billion years.

- The critical density of the universe is $3H_0^2/(8\pi G)$, or roughly $10^{-26}\,\mathrm{kg/m^3}$. The ratio of the true density of the universe to the critical density is represented by Ω. The present universe has $\Omega_{\mathrm{matter}} = 0.30$ and $\Omega_{\mathrm{dark\ energy}} = 0.70$.

Relativity

- Energy E and rest mass m are equivalent: $E = mc^2$, where c is the speed of light.

- When protons come together to create helium nuclei in the process of thermonuclear fusion, 0.7% of their mass is converted into energy.

- An observer moving by you at speed v will age $\sqrt{1 - v^2/c^2}$ years, for every year you age, according to you, where c is the speed of light. Lengths are contracted by the same factor.

- The Schwarzschild radius of a black hole of mass M is $\frac{2GM}{c^2}$. For 1 solar mass, this corresponds to a radius of about 3 kilometers.

- In special relativity, the distance between two events in spacetime is $ds^2 = dx^2 + dy^2 + dz^2 - c^2 dt^2$. In units in which $c = 1$ (e.g., distance measured in light-years and time in years), this becomes: $ds^2 = dx^2 + dy^2 + dz^2 - dt^2$.

- The Planck time is $\left(\frac{hG}{2\pi c^5}\right)^{1/2} \approx 5 \times 10^{-44}$ second.

- The Planck length is $\left(\frac{hG}{2\pi c^3}\right)^{1/2} \approx 1.6 \times 10^{-35}$ meter.

- The Planck mass is $\left(\frac{hc}{2\pi G}\right)^{1/2} \approx 2.2 \times 10^{-8}$ kilogram.

- The Einstein field equations of general relativity are $R_{\mu\nu} - \frac{1}{2}g_{\mu\nu}R = 8\pi T_{\mu\nu}$. No problems in this book involve doing calculations with this equation. But the left-hand-side involves the curvature of spacetime, and the right-hand-side involves the mass, energy, and pressure at each point in space and time.

SOLUTIONS

1. Scientific notation review

1.a 3.17×10^7.

1.b $3.14 \times 0.2 = 0.6$, to a single significant figure.

1.c 0.25 or 2.5×10^{-1}. One could round to a single significant figure, 0.3.

1.d 1×10^{-9} meters or 1.0×10^{-9} meters.

2. How long is a year?

We can almost get the right answer without a calculator. There are 60 seconds in a minute, and 60 minutes in an hour, thus $60 \times 60 = 3,600$ seconds in an hour. There are 24 hours in a day; write $24 \approx 100/4$, giving approximately $\frac{3600}{4} \times 100 = 90,000$ seconds in a day. There are 365.24 days in a year. $365 \times 9 = 365 \times 10 - 365 \approx 3,300$. But we needed to multiply by 90,000, not 9, so we need to multiply by another factor of 10^4, giving us 3.3×10^7 seconds.

If we simply multiply the numbers out, now with a calculator, we get

$$3600 \frac{\text{sec}}{\text{hour}} \times 24 \frac{\text{hour}}{\text{day}} \times 365.24 \frac{\text{days}}{\text{year}} = 3.156 \times 10^7 \frac{\text{sec}}{\text{year}},$$

which rounds to 3.2×10^7 seconds, to two significant figures. We got close doing it by hand!

3. How fast does light travel?

This is a matter of unit conversion. As we're given the speed of light to two significant figures, the final answer should have this number of significant figures:

$$c = 3.0 \times 10^8 \frac{\text{m}}{\text{sec}} \times 3.17 \times 10^7 \frac{\text{sec}}{\text{year}} \times \frac{\text{km}}{10^3 \, \text{m}} = 9.5 \times 10^{12} \frac{\text{km}}{\text{year}},$$

or about 10 trillion (10^{13}) kilometers per year. This is, of course, the distance of 1 light-year.

4. Arcseconds in a radian

Remember that there are 360 degrees, and 2π radians, in the full circle. So there are $\frac{360°}{2\pi} = \frac{180}{\pi}$ degrees per radian. There are also 3,600 arcseconds in a degree. So the number of arcseconds per radian, a number we use all the time, is

$$\frac{180}{\pi} \frac{\text{deg}}{\text{radian}} \times 3{,}600 \frac{\text{arcsec}}{\text{deg}} = \frac{18 \times 36}{\pi} \times 10^{2+2}.$$

Let's see if we can do this without a calculator:

$$36 \times 1.8 = 36 \times 2 \times 0.9 = 72 \times (1 - 0.1) = 72 - 7.2 \approx 65.$$

And by inspection, $65/\pi \approx 20$. Thus the number of arcseconds per radian is about 200,000. The precise number, using a calculator, is 206,264.8 arcseconds per radian, but the approximate value should be good enough for all calculations you will do in this problem book.

5. How far is a parsec?

Look at figure 4.1 of *Welcome to the Universe*. Consider the long skinny triangle, whose base (r) is the radius of Earth's orbit around the Sun (1 AU), whose apex angle (θ), going to the star, is 1 arcsecond, and whose height (d) is 1 parsec. The small-angle formula tells us that $\theta = r/d$, or, solving for d (which is what we're interested in here): $d = r/\theta$. This formula works if θ is in radians, but it is here in arcseconds. Luckily, we've just calculated the conversion between arcseconds and radians in problem 4, so we can plug in numbers now:

$$1\,\text{parsec} = \frac{1\,\text{AU}}{1\,\text{arcsec} \times \frac{1\,\text{radian}}{2 \times 10^5\,\text{arcsec}}} = 2 \times 10^5\,\text{AU}.$$

That is, the number of AU in a parsec is exactly the same as the number of arcseconds in a radian, by definition.

We now need this distance in light-years. We know of no clever way to do this, other than straight calculation by conversion to kilometers. We know there are 3.16×10^7 seconds in a year, from problem 2. Given the speed of light, 3×10^5 kilometers per second, the number of kilometers in a light-year is the distance light travels in 1 year (i.e., the product of these two numbers):

$$3.16 \times 10^7\,\text{sec} \times 3 \times 10^5\,\text{km/sec} = 10^{13}\,\text{km},$$

to one significant figure.

OK, now let's calculate the number of kilometers in a parsec. We know how many AU there are in a parsec, and the number of kilometers in an AU, so their product gives us

$$1\,\text{parsec} = 2 \times 10^5\,\text{AU} \times 1.5 \times 10^8\,\frac{\text{km}}{\text{AU}} = 3 \times 10^{13}\,\text{km}.$$

Comparing the two numbers, we find that 1 parsec is about 3 light-years. Using more precise values throughout, 1 parsec is 3.26 light-years.

Astronomers find themselves using parsecs and light-years for distances interchangeably, with a slight preference for parsecs. So this conversion factor of 3.26 is a useful one to keep in mind. Indeed, for rough calculations, working to a single significant figure (3 light-years per parsec) is fine. We will often find ourselves referring to kpc (kiloparsecs; the Sun is about 8 kpc from the center of the Milky Way), Mpc (megaparsecs; the Andromeda galaxy is about 0.7 Mpc from the Milky Way), and even Gpc (gigaparsecs, 10^9 parsecs; useful for talking about very distant galaxies).

6. Looking out in space and back in time

6.a We are given the distance to a single significant figure, which allows us to carry out our calculations to a single significant figure. Light travels at 3×10^8 meters per second, so the time for light to travel 10 meters is simply

$$\frac{10\,\mathrm{m}}{3 \times 10^8\,\mathrm{m/sec}} \approx 3 \times 10^{-8}\,\mathrm{sec} \times 10^9\,\frac{\mathrm{nanosec}}{\mathrm{sec}} = 30\,\mathrm{nanosec}.$$

That is, it takes light 30 nanoseconds to travel 30 feet; the speed of light is 1 foot per nanosecond (to one significant figure), a useful number to remember.

6.b This is exactly the same type of problem we've just done, and we do it in the same way. A nanometer is 10^{-9} meters, so 10 nanometers is a distance of 10^{-8} meters. Again, we're happy with a single significant figure (after all, we were not given a precise measure of the speed), so we get a time of

$$\frac{10^{-8}\,\mathrm{m}}{3 \times 10^8\,\mathrm{m/sec}} \approx 3 \times 10^{-17}\,\mathrm{sec},$$

or about a thirtieth of a femtosecond.

6.c The speed of light is 1 light-year per year, so we are seeing the Orion Nebula as it existed 1,500 years ago. One could do this the hard way, namely, converting the distance to meters and dividing by the speed of light in meters per second or meters per year, but the speed of light is especially easy to work with in units of light-years per year!

7. Looking at Neptune

7.a At this point, Neptune and Earth are on opposite sides of the Sun, and are thus 31 AU apart. We have this distance to two significant figures, so we'll work to that precision in this problem. The time for light to travel that distance is

$$t = \frac{31\,\mathrm{AU}}{3 \times 10^5\,\mathrm{km/sec}} \times 1.5 \times 10^8\,\frac{\mathrm{km}}{\mathrm{AU}} = \frac{31}{2} \times 10^3\,\mathrm{sec} = 1.55 \times 10^4\,\mathrm{sec}.$$

Note that we've written this result to three significant figures, which is more than we really should use. We are doing this in part to make sure that we can see the difference between the result we get here and that in part **b**. So we'll do the proper rounding at the end of the problem.

We are asked to convert this to hours and minutes. There are 3,600 seconds in an hour, so 15,500 seconds is 4 hours, 1,100 seconds, or about 4 hours, 20 minutes, or about $4\frac{1}{3}$ hours or 4.3 hours. Is this answer to two significant figures, as we know it should be? The answer is yes, approximately. The two significant figures are on the overall number (expressed in a single unit, such as hours), and not, for example, on the "minutes" piece of the overall number. Notice here that we might have gotten a somewhat different answer if we had rounded earlier.

7.b Again, the Sun, Earth, and Neptune lie along a straight line, but now Earth is between the Sun and Neptune. Thus the distance from Earth to Neptune is 29 AU. The same calculation then gives a time of 14,500 seconds, or almost exactly 4 hours (or 240 minutes).

8. Far, far away; long, long ago

8.a This is easy, if we remember that the speed of light is exactly 1 light-year per year. In 2017 it has been 81 years since the radio broadcast started, so the light signals have traveled 81 light-years. The sphere of radius 81 light-years, centered on us, contains tens of thousands of stars.

8.b This is just like the previous problem. Halfway to the center of the Milky Way is 13,000 light-years, rounding to two significant figures. Thus the radio signals will take 13,000 years to travel the distance.

Note that we have equated the distance from Zyborg to Sun, and from Zyborg to Earth. This is because the Earth-Sun distance (1 AU) is absolutely tiny compared with the Zyborg-Sun distance (over 10,000 light-years).

8.c Here we are interested in the round-trip time, which is the time light takes to travel $384,400 \times 2$ kilometers:

$$384,400 \, \text{km} \times 2 \times \frac{1}{3.0 \times 10^5 \, \text{km/sec}} = 2.6 \, \text{sec}.$$

Here we give the answer to two significant figures. When ground control in Houston communicated with astronauts on the Moon, this delay of 2.6 seconds was quite noticeable.

8.d The orbits of Mars and Earth are in the same plane, and they are closest when Mars is in *opposition* (i.e., when Earth lies on the line between Mars and the Sun, and is between them). Then the distance is $1.5 - 1.0 = 0.5$ AU between Mars and Earth. The time required for light to travel that distance is

$$0.5 \, \text{AU} \times 1.5 \times 10^8 \, \frac{\text{km}}{\text{AU}} \times \frac{1}{3.0 \times 10^5 \, \text{km/sec}} \times \frac{1 \, \text{min}}{60 \, \text{sec}} = 4 \, \text{min},$$

where we give our answer to one significant figure. When Mars is in *conjunction* the Sun lies between it and Earth, and Mars is then $1 + 1.5 = 2.5$ AU away, 5 times farther. This is a light travel time of 21 minutes.

This substantial delay is a real issue in communications between Earth and Mars. *Opportunity* and *Curiosity* are two robotic rovers currently exploring the surface of Mars, but they cannot be operated in real time by remote control because of this time delay. Instead, they are designed to be quite autonomous: once a day they are uploaded with detailed instructions of what to do for the day, and the rest of the time they are on their own, with software clever enough to figure out what to do if they hit an unanticipated rock.

9. Interstellar travel

9.a For constant velocity, travel time = distance/velocity, so that for $d = 4 \times 10^5$ km and $v = 4 \times 10^4$ km/h,

$$t = \frac{d}{v} = \frac{4 \times 10^5 \, \text{km}}{4 \times 10^4 \, \text{km/h}} = 10 \, \text{h}.$$

The Apollo 11 mission actually took about 3 days to reach the Moon, as the pull of Earth's gravity meant that their speed decreased throughout the trip.

9.b We know that 1 light-year is 10^{13} km, so that the distance to Proxima Centauri is $d = 4 \times 10^{13}$ km. With velocity $v = 30 \times 4 \times 10^4$ km/h, we have

$$t = \frac{d}{v} = \frac{4 \times 10^{13} \, \text{km}}{30 \times 4 \times 10^4 \, \text{km/h}}$$

$$\approx 3 \times 10^7 \text{ h.}$$

Now divide this by 24 h/day $\times$ 365 days/year $\approx 10,000$ h/year to get

$$t \approx 3,000 \text{ years.}$$

9.c For $d = 2 \times 10^6 \times 10^{13}$ km and $v = 10^3 \times 4 \times 10^4$ km/h (here we have used the fact that $30 \times 30 = 1,000$, at the level of approximation we like to use), we have:

$$t = \frac{d}{v} = \frac{2 \times 10^{19} \text{ km}}{10^3 \times 4 \times 10^4 \text{ km/h}}$$

$$\approx 5 \times 10^{11} \text{ h} \approx 5 \times 10^7 \text{ years.}$$

The speed of light is roughly 3×10^5 km/sec $\times 3600$ sec/h $\approx 10^9$ km/h. Thus the velocity we have now is roughly 4% the speed of light.

10. Traveling to the stars

10.a We are first asked to calculate the speed of Earth around the Sun. The circumference is just 2π times the radius, which we know. The time to travel that distance is 1 year, which we will need to express in seconds, so that is a useful calculation to do first:

$$1 \text{ year} \times 365 \frac{\text{days}}{\text{year}} \times 24 \frac{\text{h}}{\text{day}} \times 3,600 \frac{\text{sec}}{\text{day}}.$$

Let's see if we can do this without a calculator. We write $24 \approx 100/4$, and $365/4 \approx 90$, so

$$90 \times 100 \times 3,600.$$

$90 \times 3600 \approx 320,000$, so the final result is 3.2×10^7 seconds. The exact answer, according to a calculator (and using a year of 365.2422 days) is 3.1557×10^7 seconds, which is pretty close! We often approximate this as $\pi \times 10^7$ seconds. See also problem 2.

OK, so the speed of our orbit is then

$$\text{Speed} = \frac{\text{Distance}}{\text{Time}} = \frac{2\pi \times 1.5 \times 10^8 \text{ km}}{\pi \times 10^7 \text{ sec}} = 30 \frac{\text{km}}{\text{sec}}.$$

Let that sink in: you, and everybody on Earth, are rocketing through space, right now, at 30 kilometers every second!

10.b Our next calculation asks for the number of kilometers in 4 light-years. It will be useful to know that number for 1 light-year; multiplying by 4 at the end will be easy. We've just done the work of calculating the length of a year in seconds; the distance is that time multiplied by the speed of light:

$$\pi \times 10^7 \text{ sec} \times 3 \times 10^5 \frac{\text{km}}{\text{sec}} \approx 10^{13} \text{ km,}$$

where we've approximated that $\pi \times 3 = 10$. This is definitely a useful number: 1 light-year is 10 trillion kilometers! So 4 light years is simply 4 times that, or 4×10^{13} km.

10.c OK, traveling at 30 kilometers per second, it's going to take a long time to travel 4 light-years. We can do the arithmetic in the obvious way, but let's do it another way, just to show you a trick. (With a little bit of algebraic effort, you can save yourself a lot of work arithmetically. Remember that arithmetic is often more difficult than algebra!)

We want to calculate a travel time, given a speed (30 km/sec) and a distance (4 light-years). Let's do it this way:

$$\text{Time} = \frac{\text{Distance}}{\text{Speed}} = \frac{4\,\text{light-years}}{30\,\text{km/sec}} = 4 \times \left(\frac{3 \times 10^5\,\text{km/sec} \times 1\,\text{year}}{30\,\text{km/sec}} \right).$$

Notice that the way we wrote a light-year, we have units of km/sec on top and bottom, which cancel, leaving just years, which is what we wanted. This trick saved us the (admittedly small) amount of work it takes to calculate a time in seconds and again divide by the number of seconds in a year.

And the arithmetic is easy; the answer is 40,000 years. Another way to think about this: 30 km/sec is 10^{-4} of the speed of light, so it takes 10,000 years to travel a light-year at that speed, and 4 light-years takes 40,000 years.

11. Earth's atmosphere

11.a The density is equal to the mass of a cubic meter of air. That mass is equal to the number of molecules of N_2 times the mass per molecule. A single molecule consists of two atoms of N and thus has a mass 28 times that of hydrogen (whose mass is sometimes referred to as an atomic mass unit, or AMU). Thus

$$\text{Density of air} = 3 \times 10^{25}\,\frac{\text{molecules}}{\text{m}^3} \times \frac{28\,\text{AMU}}{\text{molecule}} \times 1.6 \times 10^{-27}\,\frac{\text{kg}}{\text{AMU}}.$$

Before you reach for that calculator, recognize that we are given the number of molecules in a cubic meter to only one significant figure, so we can do our calculation to that precision. Remember that $1.6 \approx 5/3$. The 3s cancel, leaving

$$5 \times 28 \times 10^{-2}\,\text{kg/m}^3 \approx 1.5\,\text{kg/m}^3.$$

The precise value for the density of air at room temperature at sea level is 1.225 kg/m^3; we got pretty close!

11.b The volume in question is the surface area of the Earth, $4\pi r^2$, times the thickness:

$$\text{Volume} = 8\,\text{km} \times 4 \times \pi \times (6,400\,\text{km})^2.$$

Let's try this without a calculator, again knowing that the answer need be accurate to only one significant figure (as we are given the shell thickness to a single significant figure). We have $64^2 \approx 4,000$ to a single significant figure, so we get

$$\text{Volume} \approx 8 \times 4 \times 3 \times 4 \times 10^3 \times 10^4\,\text{km}^3 \approx 4 \times 10^9\,\text{km}^3.$$

Now, the problem does not specify the units in which the results should be expressed, but in part **c** of this problem, we will want to multiply it by the density of air, which is in units of kg/m^3. So it will be useful to convert this result to cubic meters. One kilometer equals 10^3 meters, so 1 cubic kilometer equals $(10^3)^3 = 10^9$ cubic meters. So to convert, we multiply the above result by 10^9, to get: Volume $= 4 \times 10^{18}\,\text{m}^3$.

A more precise calculation of the volume of Earth's atmosphere is to calculate the spherical volume of radius $6,400 + 8$ kilometers and then subtract off the volume of radius 6,400 kilometers. We would have gotten just about the same answer and worked much harder.

11.c The first part is straightforward: the mass of an object (here, Earth's atmosphere) is the product of its volume and its density, that is, the answers to parts **a** and **b**. It is vitally important that the units match when we do such calculations!

$$\text{Mass} = \text{Density} \times \text{Volume} = 1.5\,\text{kg/m}^3 \times 4 \times 10^{18}\,\text{m}^3 = 6 \times 10^{18}\,\text{kg}.$$

With 1,000 kilograms in a ton, that is 6 quadrillion (6×10^{15}) tons, an impressive number!

We take the ratio of this to the oceans, 1.4×10^{21} kg, and find

$$\frac{6 \times 10^{18}\,\text{kg}}{1.4 \times 10^{21}\,\text{kg}} \approx 4 \times 10^{-3}.$$

Thus the atmosphere has a mass 0.4% that of the oceans.

11.d The mass of CO_2 is just the mass of the entire atmosphere multiplied by the fraction of the mass that is CO_2:

$$\text{Mass of } CO_2 = 6 \times 10^{18}\,\text{kg} \times \frac{600}{10^6} \approx 4 \times 10^{15}\,\text{kg}$$

to one significant figure. Notice how 600 ppm was turned into a fraction.

12. Movements of the Sun, Moon, and stars

12.a At the North Pole, the altitude of the Sun relative to the horizon doesn't change through a single day (see part **d**), and thus the Sun is above the horizon for 24 hours straight during the summer months (indeed, it stays above the horizon for 6 months straight!). As one heads south, the period of time, centered on the summer solstice, that the Sun stays above the horizon shrinks, until, at the Arctic Circle, latitude $+66.5°$, the Sun is above the horizon for 24 hours on the summer solstice (June 21) but sets below the horizon at some point on all other days of the year. Similarly, at the South Pole, one has 6 months of sunshine centered on December 21, and at a latitude of $-66.5°$, December 21 is the only day with 24 hours of sunshine.

12.b If the Moon is half-illuminated by the Sun, then the Sun, Earth, and Moon form a right angle; that is, the Sun and Moon will be about 90 degrees apart as seen by an observer on Earth. The problem didn't specify whether this is a first-quarter or third-quarter Moon, so this happens twice every month, for roughly 6 hours (the time between sunrise and moonset, or between moonrise and sunset, depending on whether it is first or third quarter). Note that the half-illuminated Moon is quite visible during the day; it is plenty bright enough to be seen against the bright daytime sky.

12.c The orbit of Mars lies outside the orbit of Earth. Thus from our perspective, it never happens that Mars appears to pass in front of the Sun.

12.d If one stands at the Earth's North Pole, Polaris will appear directly overhead and all the other stars will appear to rotate around in circles parallel to the horizon, centered on Polaris. Hence, they will never set. Similarly at Earth's South Pole, stars appear to rotate in circles parallel to the horizon, around a point straight up (where there does not happen to be a bright star to mark the South Celestial Pole).

12.e To a very good approximation, Polaris lies directly along the line extending North from the axis of the Earth (it is less than 1 degree away from the true North Pole). Therefore it appears directly overhead from the North Pole, essentially without appearing to move during the entire 6 months of winter darkness. Note that the precession of Earth's axis causes the direction of the North Pole to circle relative to the background stars on a timescale of 26,000 years; thus Polaris will not be as close to the North Pole thousands of years from now.

13. Looking at the Moon

13.a We know that when the Moon is full, the Sun, Earth, and Moon line up in an approximate straight line. Thus if the Moon is on the eastern horizon (i.e., just rising), the Sun is on the western horizon (i.e., just setting). That is, it is early evening, roughly 6 p.m.

13.b Venus' orbit around the Sun is inside that of Earth. Thus, Venus can never be found too far away from the Sun (a maximum of 47°, as it turns out, which you can calculate yourself knowing a bit of trigonometry and the fact that Venus' orbit has a radius of 0.7 AU). It therefore is never seen opposite the Sun; it could not be rising while the Sun is setting. The planet must be Jupiter.

13.c The Moon rotates around its own axis at the same rate it orbits Earth (it is "tidally locked"), and thus it shows the same face to Earth all the time. Thus if you live on the far side, you will never see Earth. You will, however, see the Sun. You rotate around your axis once a month, and the Sun will therefore rise roughly once per month; that is, 1 "Moon-day" is roughly 30 Earth-days.

14. Rising and setting

14.a No, stars do not rise and set for an observer sitting at the North Pole. Over the space of 24 hours, stars appear to circle around the North Pole of the sky. The North Pole of the sky is directly overhead as seen from an observer at Earth's North Pole, so stars move in circles parallel to the horizon (i.e., at constant elevation above the horizon) and thus never set.

14.b For an observer at the North Pole, the Sun, like all the other stars, moves parallel to the horizon over 24 hours and therefore does not appear to rise or set. Over the course of the year, however, the Sun moves in the sky relative to the background stars, and thus changes its elevation. In particular, it rises once (on the Vernal Equinox, March 21) for a 6-month-long "day," and it sets for a cold and dark 6-month-long "night" on the autumnal equinox, September 21. Since Alaska is close to the North Pole, the Sun stays close to constant elevation through 24 hours. Therefore, it stays bright even at midnight at the height of summer.

There is an imaginary line, the Arctic Circle, at +66.5° latitude, north of which there is a period of time during the year when the Sun is indeed above the horizon for a full 24 hours. Even if you are somewhat south of that line (which most of Alaska is), the Sun doesn't get very far below the horizon at midnight at the height of summer, so it never gets very dark. All of this happens the same way at the South Pole, but with the seasons reversed on the calendar (thus the summer solstice is on December 21).

14.c As Polaris is (almost) in line with the axis of rotation, and the axis of rotation passes vertically through the ground, Polaris will be exactly overhead as seen from the North Pole. And as the direction of the axis of rotation of Earth does not change, Polaris will remain there all year long.

On long timescales, however, North Pole actually does move, or "precess." Due to torques from both the Moon and the Sun, the direction of Earth's spin axis moves through the sky at a rate of 1° every 175 years, making a full circle on the sky (with a radius of 23.5°) in 26,000 years. The effect is large enough to be (barely) observable over a human lifetime without the use of telescopes, and in fact was known to the ancient Greeks.

14.d If you are at the equator, the axis of rotation is parallel to the ground, that is, it is horizontal. Therefore Polaris appears due north, right on the horizon (and thus would be visible only to observers with nothing obscuring the horizon). Keep in mind that Polaris is many light-years away, so the fact that we sit on the surface of Earth, not exactly on its axis 6,400 km away, makes a completely negligible difference in angle.

As in part **c**, the axis of rotation of Earth does not change with time, so the position of Polaris also does not change.

14.e See the accompanying figure 8. We know that the position of Earth's axis does not change, as seen from the poles, from the equator, or even from Princeton; thus Polaris will remain at the same place in the sky as seen from any position on Earth in the Northern Hemisphere (it is of course always invisible, below the horizon, to observers in the Southern Hemisphere). Looking from Princeton (latitude +40°), its position is 40° above the horizon due north (indeed, this is one of the standard ways of deducing one's latitude in the Northern Hemisphere). As at the North Pole (part **a**), the stars will appear to travel in circles centered on Polaris.

The diagram shows that all the stars that are less than 40° away from Polaris will be always closer to Polaris than the horizon is, and therefore will not set. Stars

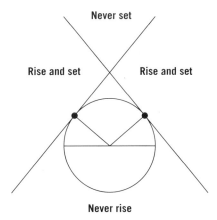

Figure 8. Diagram for problem 14.e. Dots show the positions of Princeton on Earth's surface at different times as Earth rotates about its axis (vertical toward Polaris). The straight lines are the local horizons at each point; we can only see stars above the horizon. The cone at the top is where stars do not set as seen from Princeton.

farther than 40° will, at some point or another, be farther away than the horizon and thus will appear to rise and set. Indeed, stars more than 140° from Polaris will never appear above the horizon from Princeton; you would have to travel to lower latitudes to see them.

15. Objects in the sky

15.a If the Moon is high in the sky, it is not close to either rising or setting. At full Moon, the Sun and the Moon are opposite, so the full Moon rises as the Sun sets (6 p.m.) and sets as the Sun rises (6 a.m.). If the Moon is high in the sky, we're roughly half-way between the two times, that is, about midnight.

15.b No. Over a 24-hour period, all objects describe circles around Polaris. On the equator, Polaris is right on the horizon, so all circles going around it must cross the horizon (i.e., rise and set). No objects are circumpolar as seen from the equator.

15.c Yes. Polaris is directly overhead, and all objects describe circles parallel to the horizon; they don't rise or set over 24 hours. So if the Moon is visible at all, it will be circumpolar.

Note that it is not true that the Moon is always visible from the North Pole. It rises (comes above the horizon) once per month and sets once per month, because its orbit is tilted relative to Earth's equator.

15.d If the Moon is in Earth's shadow, it is on the opposite side of the sky from the Sun and therefore is full. At midnight, it is therefore high in the sky. A lunar eclipse lasts for a few hours: you first see a fully illuminated Moon, then as it starts to slip into Earth's shadow, ever more of its disk appears dark. During a total lunar eclipse, the Moon is dimly illuminated by sunlight refracted through Earth's atmosphere, and thus, like a sunset, appears quite red.

16. Aristarchus and the Moon

16.a The trick to this question is to remember that the angles AES and EMS are both right angles (as is stated in the problem itself). First, consider the triangle spanned by the points M, E, and S. The angles of this triangle add up to 180°:

$$\angle \text{EMS} + \angle \text{ESM} + \angle \text{SEM} = 180°.$$

But, because $\angle$EMS is a right angle (i.e., 90°), this simplifies to

$$\angle \text{ESM} + \angle \text{SEM} = 90°.$$

Next, consider the angle AES. Consideration of the figure shows that

$$\angle \text{AEM} + \angle \text{SEM} = \angle \text{AES}.$$

But $\angle$AES is a right angle, and therefore

$$\angle \text{AEM} + \angle \text{SEM} = 90°.$$

Consideration of this equation, and the one four lines above, immediately shows that

$$\angle \text{AEM} = \angle \text{ESM},$$

which is what we were requested to prove.

16.b We apply the small-angle formula to the angle $\angle \text{ESM}$; the distance $\overline{\text{EM}}$ subtends the angle $\angle \text{ESM}$ at a distance $\overline{\text{ES}}$, so the small-angle formula states simply:

$$\angle \text{ESM} = \frac{\overline{\text{EM}}}{\overline{\text{ES}}} \text{ (in radians)}.$$

The distances from Earth to the Moon and from Earth to the Sun, are 4×10^5 kilometers, and 1.5×10^8 kilometers, respectively. Therefore

$$\angle \text{ESM} = \frac{4 \times 10^5 \,\text{km}}{15 \times 10^7 \,\text{km}} \approx \frac{1}{400} = 0.0025 \,\text{radians}.$$

To convert to degrees, we multiply by $\frac{180°}{\pi \,\text{rad}} \approx 60$, to get $0.16° \approx 10$ arcminutes. Thus the angle SEM, between the first-quarter Moon and the Sun, is this much less than $90°$, that is, $89.84°$.

16.c The small-angle formula relates the physical diameter s of an object, its distance d, and its angular diameter θ: $s = \theta d$. Writing this expression down for both the Moon and the Sun, and taking the ratio, we find:

$$\frac{s_{\text{Moon}}}{s_{\text{Sun}}} = \frac{\theta_{\text{Moon}} d_{\text{Moon}}}{\theta_{\text{Sun}} d_{\text{Sun}}}.$$

However, the angular diameters of the Sun and the Moon are the same (an amazing coincidence, which is why the Moon can cover the disk of the Sun essentially perfectly to produce a solar eclipse), and so the θs drop out of the ratio. Thus the ratio of angular sizes is equal to the ratio of distances. And we have seen above that the ratio of the Moon distance to the Sun distance is equal to the angle AEM, expressed in radians.

Aristarchus determined that that angle was $3°$ (quite a bit larger than the true value, which you calculated in part **b** as $0.16°$). $3°$ is equivalent to $\frac{1}{20}$ radians. Thus he concluded that the ratio of the size of the Sun to the size of the Moon was 20. He also estimated that the Moon's diameter was $\frac{1}{3}$ the diameter of Earth (pretty close to the modern value of 0.27) by comparing the shape of Earth's shadow on the Moon during a lunar eclipse to the Moon's limb. He thus concluded that the Sun had a diameter about 7 times that of Earth. Again, the numerical value is wrong; in reality, the Sun has a diameter 109 times that of Earth, but the qualitative statement that the Sun is much larger than Earth was correct. Just as the smaller Moon orbits the larger Earth, Aristarchus concluded, almost two millennia before Copernicus, that Earth orbits the Sun.

17. The distance to Mars

This is a parallax question. Consider a long skinny triangle with its apex at Mars, an interior angle of $\alpha = 2$ arcseconds, and a base of $s = 1,000$ kilometers. We want the distance d; the small-angle formula tells us that

$$d = \frac{s}{\alpha},$$

where we need to express α in radians. There are 200,000 arcseconds in a radian, so this angle is 10^{-5} radians. Thus the distance is

$$d = \frac{1,000 \,\text{km}}{10^{-5}} = 10^8 \,\text{km}.$$

This is a bit under 1 AU. Notice that the distance between Mars and Earth changes as the two planets move in their respective orbits. The closest they ever get is about 0.5 AU. This happens when Mars is at *opposition*, that is, when Earth lies directly between Mars and the Sun, and thus Mars is directly opposite the Sun in the sky, as seen from Earth.

18. The distance to the Moon

Consider a triangle with its apex at the Moon, and the two long sides going to you and your friend. The Moon is directly overhead, so that the line between you and your friend is perpendicular to the height of the triangle. The distance to the Moon is 400,000 kilometers, so the small angle formula gives the angle, in radians, as

$$\theta = \frac{200\,\text{km}}{400,000\,\text{km}}.$$

Before multiplying this out, let's convert right away to arcseconds, by multiplying by the number of arcseconds in a radian, 200,000:

$$\theta = \frac{200}{400,000} \times 200,000\,\frac{\text{arcsec}}{\text{radian}} = \frac{200}{2}\,\text{arcsec} = 100''.$$

The positions differ by 100 arcseconds.

Incidentally, you had to synchronize your watches to make sure you made the measurement of the Moon's positions at the same time. The Moon is in orbit around Earth, and thus its position relative to the background stars is constantly changing. You do not want to confuse the parallax measurement with this effect!

19. Masses and densities in the solar system

19.a The small-angle formula states that the angle in radians an object subtends on the sky is its size divided by its distance to us. The object's size refers to its diameter, and we want its radius, so the radius of the Sun subtends an angle of $\theta = 0.25°$. Thus the small-angle formula says that the radius of the Sun is its distance times this angle (converted to radians):

$$R_{\text{Sun}} = d \times \theta \approx 1.5 \times 10^8\,\text{km} \times 0.25° \times \frac{1\,\text{radian}}{60°} = \frac{2.5 \times 10^7\,\text{km}}{40}$$

$$= \frac{5}{8} \times 10^6\,\text{km} \approx 6.3 \times 10^5\,\text{km}.$$

Here we used an approximate conversion from radians to degrees. To calculate the density of the Sun, we need to divide its mass by its volume. The mass is $M_{\text{Sun}} = 3 \times 10^5 \times M_{\text{Earth}}$, the volume is that of a sphere with radius R_{Sun}. The density ρ is therefore

$$\rho_{\text{Sun}} = \frac{M_{\text{Sun}}}{\frac{4\pi}{3} R_{\text{Sun}}^3} \approx \frac{3 \times 10^5 \times 6 \times 10^{24}\,\text{kg}}{4 \times (6.3 \times 10^5\,\text{km} \times 10^3\,\text{m/km})^3} \approx \frac{2 \times 10^{30}\,\text{kg}}{4 \times 250 \times 10^{24}\,\text{m}^3}$$

$$= 2 \times 10^3\,\frac{\text{kg}}{\text{m}^3} = 2\rho_{\text{water}}.$$

The Sun has about twice the density of water. *Note: The correct answer is closer to 1.4 times the density of water; the biggest error came from our rather approximate value for the angular size of the Sun; the correct value is about 0.55°.*

19.b The mass of the solar system is dominated by the Sun, as we discuss in part **d**. To calculate the density of the solar system we need to find the volume of a sphere with radius 30 AU.

$$\rho_{\text{solar system}} \approx \frac{M_{\text{Sun}}}{\frac{4\pi}{3}R_{\text{Neptune}}^3} \approx \frac{2 \times 10^{30}\,\text{kg}}{4 \times (30 \times 1.5 \times 10^8\,\text{km} \times 10^3\,\text{m/km})^3}$$

$$\approx \frac{2 \times 10^{30}\,\text{kg}}{4 \times 90 \times 10^{36}\,\text{m}^3} \approx 5 \times 10^{-9}\,\frac{\text{kg}}{\text{m}^3} \approx 5 \times 10^{-12}\,\rho_{\text{water}}.$$

19.c The comets contribute almost no mass. Therefore, by analogy with part **b**, we obtain the average density out to the Oort cloud by dividing the solar mass by the volume of a sphere with radius 1 light-year:

$$\rho \approx \frac{M_{\text{Sun}}}{\frac{4\pi}{3}R_{\text{Oort}}^3} \approx \frac{2 \times 10^{30}\,\text{kg}}{4 \times \left(1\,\text{ly} \times \frac{10^{16}\,\text{m}}{1\,\text{ly}}\right)^3} \approx \frac{2 \times 10^{30}\,\text{kg}}{4 \times 10^{48}\,\text{m}^3} \approx 5 \times 10^{-19}\,\frac{\text{kg}}{\text{m}^3} \approx 5 \times 10^{-22}\,\rho_{\text{water}}.$$

This is a very low density!

19.d The most massive object in the solar system after the Sun is Jupiter, with a mass of only 1/1000 that of the Sun. The sum of the masses of all the other planets is less than 1.5 times that of Jupiter (a total of 447 Earth masses, or 0.15% the mass of the Sun), so the approximation that the mass of the planets is negligible in the solar system is a very good one.

20. Forces on a book

20.a The book appears to be at rest relative to our surroundings. Not only that, but it is remaining at rest, so its velocity is not changing; the acceleration must be zero. Newton's second law tells us that the sum of the forces on the book is equal to the mass of the book times its acceleration, which is zero in this case. Therefore, the sum of all the forces on the book is equal to zero. There is certainly the gravitational force of Earth downward upon the book, and therefore an equal and opposite force must exist to balance this out. This force is applied by the table. The book pushes down on the table with the force of its weight, and the table pushes back with the same force by Newton's third law, and therefore the forces on the book are balanced, as they must be.

This upward force from the table on the book, which keeps it from being accelerated toward the floor, is often called the contact force, or normal force. The table, being rigid, will supply such a contact force for an arbitrarily large weight, until the table eventually collapses (no object is infinitely rigid!).

But wait, let us look at this question again. We know that the book is moving in a circle, because Earth is rotating around its axis. Is the book orbiting Earth? No, it is clearly not; orbital motion happens when gravity is the only force acting. But the book does have a velocity relative to Earth's central axis; moreover, the velocity is changing (Earth turns at a constant rate, but the direction of motion of the book is constantly changing). For definiteness, let us put this book on a table in Quito, Ecuador, on Earth's equator. How large is the acceleration? For uniform circular motion at speed v in a circle of radius r, the acceleration a is $a = v^2/r$. How large is v? The book makes a full circuit (distance $2\pi r$, where r is the radius of Earth) in $P = 24$ hours, or roughly 10^5 seconds. The speed v is thus $2\pi r/P$, and thus the acceleration is

$$a = \frac{v^2}{r} = \frac{4\pi^2 r}{P^2} \approx \frac{4\pi^2 \times 6 \times 10^6 \,\mathrm{m}}{(10^5 \,\mathrm{sec})^2} \approx 2.5 \times 10^{-2} \,\mathrm{m/sec^2}.$$

(Notice the useful approximation $\pi^2 \approx 10$.) This is a small acceleration, a few tenths of a percent of the acceleration of gravity due to Earth on Earth's surface (about 10 m/sec^2).

So the book does have a net acceleration, albeit a small one. The contact force of the table on the book does not exactly cancel the force of Earth's gravity on the book, because the two together must add up to the mass of the book times the acceleration that we've just calculated. For most everyday problems, it is perfectly acceptable to ignore this very tiny net force that keeps the book going in a circle; however, we went through this argument to explain in detail how Newton's second law can be applied.

20.b Well, as long as we're worrying about the acceleration that stems from the book going around Earth due to Earth's rotation, should we also think about the fact that Earth, with the table and book on it, is going around the Sun? We should! After all, this motion is also circular, and therefore is not motion at constant velocity. Let us work out this acceleration:

$$a = \frac{4\pi^2 r}{P^2} \approx \frac{4\pi^2 \times 1.5 \times 10^{11} \,\mathrm{m}}{(3 \times 10^7 \,\mathrm{sec})^2} \approx 6 \times 10^{-3} \,\mathrm{m/sec^2},$$

where we plugged in the radius and period of Earth's orbit around the Sun. This acceleration is due to the gravitational force of the Sun on the book. The acceleration is smaller than that due to Earth's rotation, but only by a factor of 4 or so. So the total acceleration on the book is the vector sum of the two accelerations we've just calculated.

The acceleration of the book due to its motion around the Sun may seem small, but it would be just right to keep the book in orbit around the Sun, if we were magically able to take away Earth and the table, leaving the book alone floating in space.

20.c This problem asks for the ratio of the strength of the gravitational force from all the planets and the Moon acting on the book, to the force from the Sun. We're putting all the planets in a line, extending from Earth in the direction of the Sun. This makes it easy to calculate the distance of each from Earth. Note that the outer planets are on the opposite side of the Sun relative to Earth, and in order of distance from Earth, we have the Moon, Venus, Mercury, Sun, Mars, Jupiter, Saturn, Uranus, and Neptune.

The gravitational force on the book from each body is

$$F_{\mathrm{planet}} = \frac{GM_{\mathrm{planet}}M_{\mathrm{book}}}{r^2},$$

where r is the distance from the book to the planet in question. The force on the book from the Sun is given by the same expression (with the mass of the Sun and distance to the Sun inserted, of course). Therefore, the ratio of the force on the book from the Sun, and from any given planet, is

$$\frac{F_{\text{planet}}}{F_{\text{Sun}}} = \frac{M_{\text{planet}}}{M_{\text{Sun}}} \left(\frac{1\,\text{AU}}{r_{\text{planet}}} \right)^2.$$

Again, r here refers to the distance between the book (i.e., Earth; the planets are far enough away to allow us to approximate Earth as infinitely small for this calculation) and the planet.

In the following table, we tabulate this ratio for each of the planets and the Moon, using the masses of the planets given in table 9.1 of *Welcome to the Universe*:

Planet	r (AU)	$M_{\text{planet}}/M_{\text{Sun}}$	$F_{\text{planet}}/F_{\text{Sun}}$
Moon	0.00252	3.7×10^{-8}	5.9×10^{-3}
Venus	0.28	2.7×10^{-6}	3.4×10^{-5}
Mercury	0.61	1.7×10^{-7}	4.6×10^{-7}
Mars	2.52	3.3×10^{-7}	5.2×10^{-8}
Jupiter	6.2	1.0×10^{-3}	2.6×10^{-5}
Saturn	11	2.9×10^{-4}	2.4×10^{-6}
Uranus	20	4.4×10^{-5}	1.1×10^{-7}
Neptune	31	5.1×10^{-5}	5.3×10^{-8}

The planets and the Moon are all in the same direction of the sky, so their gravitational pulls add directly. We find the ratio of the total force on the book from these planets to the pull from the Sun to be 6.0×10^{-3}, which is truly tiny (the Moon, being closest, contributes the most). This is not surprising; the motions of objects in the Solar system are dominated by the Sun.

Thus the Moon and the other planets contribute only a small component of the force on the book. Similarly, the forces of the planets on each other are small compared to the force of the Sun on each of them. Nevertheless, these forces are nonzero and have a measurable effect on the planetary orbits. They must be taken into account if you want to understand the orbits of the planets in detail. Because of these additional forces, the orbits of the planets are not perfect ellipses. The calculation of these refinements to the orbits is now done with sophisticated computer programs.

21. Going ballistic

21.a This is straightforward; the satellite is traveling in a circular orbit of radius $400 + 6,400 = 6,800$ km around Earth's center. The circumference is 2π times this, and it travels this distance in 90 minutes, or 5,400 seconds. Its speed thus is

$$\text{Speed} = \frac{\text{Distance}}{\text{Time}} = \frac{2\pi \times 6,800\,\text{km}}{5,400\,\text{sec}} \approx 8\,\frac{\text{km}}{\text{sec}}.$$

(We used a calculator to do the arithmetic here.) If we want the answer in kilometers per hour, we can simply multiply the above by 3,600 seconds in an hour, to get

$$8\,\frac{\text{km}}{\text{sec}} \times 3,600\,\frac{\text{sec}}{\text{hour}} \approx 30,000\,\frac{\text{km}}{\text{hour}}.$$

21.b This is easiest done by writing down the relationship between the orbital velocity v, the mass of the central gravitating object M (Earth in this case) and the radius of the orbit r:

$$v^2 = \frac{GM}{r}.$$

If we solve for M (which is what we're trying to find), we get

$$M = \frac{v^2 r}{G}.$$

So now it is a matter of plugging in numbers, being very careful to be consistent about units (let us convert everything to MKS). We find:

$$M_{\text{Earth}} = \frac{(8 \times 10^3 \, \text{m/sec})^2 \times 6.8 \times 10^6 \, \text{m}}{6.67 \times 10^{-11} \, \text{m}^3/\text{kg/sec}^2}.$$

Note the funny units associated with G, but the end result will indeed be in kilograms, as it should be. If we approximate $6.8/6.67 = 1$, we find:

$$M_{\text{Earth}} = 6.4 \times 10^{24} \, \text{kilograms},$$

or about 6×10^{21} tons.

22. Escaping Earth's gravity?

Let the distance between Earth and the Moon be D, and the distance of a satellite along that line from Earth be r. We'll do this problem algebraically as long as we can before plugging in numbers: algebra is usually easier than arithmetic! We need the distance r where the gravitational force from Earth and the Moon are equal:

$$F = \frac{Gm M_{\text{Earth}}}{r^2} = \frac{Gm M_{\text{Moon}}}{(D-r)^2},$$

where m is the mass of the satellite. We can simplify this a bit by canceling some common factors and rearranging to find:

$$\left(\frac{D-r}{r} \right)^2 = \frac{M_{\text{Moon}}}{M_{\text{Earth}}}.$$

Take the square root of both sides. On the right-hand side, we have the square root of the ratio of the mass of the Moon and Earth. The Moon has a mass of about 1% the mass of Earth; the square root of this ratio is 0.1. So we have:

$$\frac{D-r}{r} = 0.1$$

$$D - r = 0.1r$$

$$1.1r = D$$

$$r = \frac{D}{1.1} \approx 0.9 \, D.$$

That is, about 90% of the way between Earth and the Moon (about 340,000 kilometers from Earth), the gravitational pull of the two bodies is the same.

Here's something to think about: when you took the square root above, there were two solutions, a positive one (which is what we used) and a negative one. What does the negative solution correspond to, physically? Let's work this out: following the arithmetic above, but adding a negative sign, we find:

$$-(D - r) = 0.1r$$

$$r = \frac{D}{0.9} \approx 1.1\,D.$$

Here, the spacecraft is farther from Earth than the Moon is. There is a point on the other side of the Moon where gravity on the spacecraft from both Earth and the Moon are the same, but in this case, they are pointing in the same, not opposite directions.

When the Apollo astronauts first went to the Moon, it was often erroneously stated that after they passed the point where the Moon's gravity was stronger than Earth's gravity, they had "broken free of the Earth's gravity." This is not true, of course; they still feel the GmM_{Earth}/r^2 force from Earth at any distance from Earth.

23. Geosynchronous orbits

This is another application of Kepler's third law; we've got a period (24 hours) and want to find the radius of the orbit. However, note here that this is not an orbit around the Sun, and so Kepler's third law in its original form is not valid. Rather, we can use Newton's form of Kepler's third law:

$$a^3 = \frac{GM_\oplus P^2}{4\pi^2}.$$

In doing this calculation, we need to work in MKS units throughout. $M_\oplus = 6 \times 10^{24}$ kg is the mass of Earth, $G = \frac{2}{3} \times 10^{-10}\,\text{m}^3\,\text{sec}^{-2}\,\text{kg}^{-1}$ is Newton's constant, and the period is 1 day, which we'll approximate as 90,000 seconds. Thus,

$$a^3 = \frac{\frac{2}{3} \times 10^{-10}\,\text{m}^3\,\text{sec}^{-2}\,\text{kg}^{-1} \times 6 \times 10^{24}\,\text{kg} \times 8 \times 10^9\,\text{sec}^2}{40},$$

where we've approximated $\pi^2 \approx 10$. Gathering the terms together, we find:

$$a^3 = 8 \times 10^{22}\,\text{m}^3.$$

So far, no calculator. How are we going to take the cube root without a calculator? Well, we know that $8 \times 10^{22} = 80 \times 10^{21}$, and the cube root of 10^{21} is 10^7. We also know that $4^3 = 64$ and $5^3 = 125$, so the cube root of 80 is a shade over 4. So to a single significant figure, we find

$$a = 4 \times 10^7\,\text{m} = 40{,}000\,\text{km}.$$

Are we done? Well, we were asked for the distance from Earth's surface, whereas what we've calculated is from Earth's center. So we need to subtract from this the radius of Earth, 6,400 km, leaving roughly 34,000 km. And we're asked to express this number in Earth radii: dividing by 6,400 km gives a bit over 5 Earth radii.

If we redo the calculation using precise values for Earth's mass, G, and π (and using a calculator), we find a distance of 35,900 km above Earth's surface, about 5.63 Earth radii.

Another way to do the problem would be to recognize that Kepler's third law can be written in the form

$$a^3 = P^2/M,$$

with a in AU, P in years, and M in solar masses. This form has the advantage of skipping the need to look up the values of G and π, but it would require the additional work of converting all our numbers to those (in this case inconvenient) units.

There's a third way to do it, which is our favorite. We know that satellites in low-Earth orbit have a period of about 90 minutes. The radius of that orbit is the radius of Earth, 6,400 kilometers. Kepler's third law says that $a \propto P^{2/3}$ for orbits around a common body, so

$$\frac{a_{\text{geosynchronous}}}{6,400 \, \text{km}} = \left(\frac{24 \, \text{h}}{90 \, \text{min}}\right)^{2/3} = 16^{2/3}.$$

Here, we are reduced to using a calculator:

$$a_{\text{geosynchronous}} = 6.3 \times 6,400 \, \text{km} = 40,000 \, \text{km},$$

to one significant figure.

24. Centripetal acceleration and kinetic energy in Earth orbit

24.a Let's do it both ways. We know that the period is related to the speed of the orbit by $P = 2\pi r/v$. Putting this into Kepler's third law yields

$$r^3 = \frac{4\pi^2 G M_\oplus r^2}{4\pi^2 v^2}.$$

Canceling the common $4\pi^2$ term and solving for v gives

$$v = \sqrt{\frac{GM_\oplus}{r}}.$$

Alternatively, let's argue from Newton's laws. The force on the satellite of mass m is given by Newton's law of gravity, $GM_\oplus m/r^2$. It is equal to its mass times its acceleration; the latter is given by the standard expression for circular motion, v^2/r. Equating the two gives

$$m\frac{v^2}{r} = \frac{GM_\oplus m}{r^2}.$$

Canceling the common factor of m and solving for v again gives

$$v = \sqrt{\frac{GM_\oplus}{r}}.$$

24.b We can use the equation just derived to determine the speed. The one tricky bit is to figure out what radius we want to use. The height of the satellite above Earth's surface is 850 km, but what counts here is the distance from the center of Earth, 6,400 km away. Thus we use $r = (6,400 + 850) \, \text{km} \times 10^3 \, \text{m/km} = 7 \times 10^6 \, \text{m}$, to a single significant figure. Putting in numbers gives

$$v = \sqrt{\frac{GM_\oplus}{r}} = \sqrt{\frac{\frac{2}{3} \times 10^{-10} \text{m}^3 \sec^{-2} \text{kg}^{-1} \times 6 \times 10^{24} \, \text{kg}}{7 \times 10^6 \, \text{m}}} \approx \sqrt{\frac{1}{2} \times 10^8 \, \frac{\text{m}^2}{\sec^2}} \approx 7,000 \, \text{m/sec}.$$

The satellite was moving at 7 kilometers per second; that's fast!

What is the period? We could return to Kepler's third law for that, but a more straightforward way to do this calculation is to remember that the period is the distance the satellite travels in one orbit, divided by the speed:

$$P = \frac{2\pi r}{v} = \frac{2\pi \times 7 \times 10^6 \, \text{m}}{7 \times 10^3 \, \text{m/sec}} \approx 6,000 \, \text{sec},$$

or about 100 minutes. A satellite in low-Earth orbit goes once around in about an hour and a half (the precise value depending on the exact altitude above Earth's surface).

24.c This problem is straightforward; the mass is the volume times the density. We just have to be careful with the units as we calculate:

$$m = V\rho = 10\,\text{cm} \times 10\,\text{cm} \times 2\,\text{mm} \times \frac{1\,\text{cm}}{10\,\text{mm}} \times 2.7\frac{\text{g}}{\text{cm}^3} \times \frac{1\,\text{kg}}{10^3\,\text{g}} \approx 5 \times 10^{-2}\,\text{kg}.$$

That is, about 50 grams, or 2 ounces, to a single significant figure (given the rough dimensions, more precision is not appropriate here).

24.d The kinetic energy of an object of mass m moving at speed v is $KE = \frac{1}{2}mv^2$. We calculated the mass in part **c** and the speed in part **b**, so this is straightforward:

$$KE = 0.5 \times 5 \times 10^{-2}\,\text{kg} \times (7 \times 10^3\,\text{m/sec})^2 \approx 10^6\,\text{Joules},$$

to a single significant figure. The next problem will put this amount of energy in context.

24.e The problem gives us a hint, by suggesting that we calculate the ratio of energies. Often in taking ratios, some numbers will drop out, which will save us some arithmetic. The kinetic energy of an object of mass m moving at speed v is $\frac{1}{2}mv^2$ (this is the energy of a collision), while the explosive energy of the same mass of TNT is mp, where $p = 4.2 \times 10^6$ Joules per kilogram is the explosive energy per kilogram. The ratio of the two is thus:

$$\frac{\frac{1}{2}mv^2}{mp} = \frac{1}{2}\frac{v^2}{p}.$$

Plugging in numbers (with v from part **b**), and being careful with our units, we find

$$\frac{1}{2}\frac{(7 \times 10^3\,\text{m/sec})^2}{4.2 \times 10^6\,\text{kg}\,\text{m}^2/\text{sec}^2/\text{kg}} = \frac{0.5 \times 5 \times 10^7}{4.2 \times 10^6} \approx 6.$$

First, note that the mass dropped out; the answer here did not depend on that in part **c**. Second, notice the work done on the units; we plugged in the definition of a Joule, and by the end, all the units dropped out (as they had to for a meaningful ratio). Finally, marvel at the answer: a collision from a piece of space debris releases about 6 times more energy than an equivalent amount of TNT! No wonder people are so concerned about space debris!

The problem is not quite as bad as we've calculated, simply because if you're in orbit, you will also be traveling at 7 km/sec, and what counts in the energy of collision is the relative speed of you and the piece of debris. As long as you travel in the same direction, the collision is more gentle than you've just calculated.

Space debris is a real problem in low-Earth orbit; there are enough satellites that collisions between them is a real hazard. Moreover, if a satellite is broken into many pieces in such a collision, those pieces now become space junk themselves, making the problem worse. The Hollywood movie *Gravity* illustrates this rather dramatically, as the Hubble Space Telescope is destroyed by the debris of a damaged satellite.

25. Centripetal acceleration of the Moon and the law of universal gravitation

25.a The distance from Earth to the Moon can be expressed in terms of Earth radii by calculating a ratio:

$$\frac{\text{Distance from Earth to Moon}}{\text{Radius of Earth}} = \frac{384,000\,\text{km}}{6,400\,\text{km}} = 60.$$

To calculate a speed for the Moon's orbit, we ask how far it travels in 1 month. It goes in a circle whose circumference is $2\pi r$, and the speed is the distance (i.e., this circumference) divided by the time ($P = 27.3$ days). Plugging everything in, and being careful with the units yields

$$v = \frac{2\pi r}{P} = \frac{2\pi \times 3.84 \times 10^8\,\text{m}}{27.3\,\text{days} \times 86,400\,\text{sec/day}} = 1.02 \times 10^3\,\frac{\text{m}}{\text{sec}}.$$

The centripetal acceleration for a body moving in a circle of radius r at speed v is v^2/r, so the acceleration is

$$\text{Acceleration} = \frac{v^2}{r} = \frac{(1,020\,\text{m/sec})^2}{3.84 \times 10^8\,\text{m}} \approx 2.7 \times 10^{-3}\,\frac{\text{m}}{\text{sec}^2}.$$

25.b First, remember that we can express the period of the orbit in terms of its circumference, $2\pi r$, and its speed v:

$$P = \frac{2\pi r}{v}.$$

Thus Kepler's third law states

$$r^3 \propto P^2,$$

or

$$r^3 \propto \left(\frac{2\pi r}{v}\right)^2.$$

A little rearrangement gives

$$v^2 \propto \frac{1}{r}.$$

The force is proportional to the acceleration, $\frac{v^2}{r}$; that is,

$$F = m\frac{v^2}{r} \propto \left(\frac{1}{r}\right)\left(\frac{1}{r}\right) = \frac{1}{r^2}.$$

Thus Newton demonstrated that the relationship between acceleration and force, combined with Kepler's third law, leads to an inverse square law for the force of gravity.

25.c We found in part **b** that the acceleration of objects falling to Earth is proportional to the inverse square of the distance from Earth's center:

$$\frac{\text{Acceleration at Earth's surface}}{\text{Acceleration at Moon}} = \frac{(\text{Distance to Moon})^2}{(\text{Radius of Earth})^2} = \left(\frac{\text{Distance to Moon}}{\text{Radius of Earth}}\right)^2.$$

But the ratio inside the parentheses is one we just calculated in part **a**, namely 60. Thus, if Newton was right, and gravity from Earth is responsible for the acceleration of both the Moon and an apple falling from a tree, then the apple's acceleration should be $(60)^2 = 3,600$ times larger than that of the Moon. Multiplying the Moon's acceleration, which we calculated, by 3,600, gives us

Expected acceleration on Earth's surface $= 2.7 \times 10^{-3}\,\mathrm{m/sec^2} \times 3600 = 9.75\,\mathrm{m/sec^2}$.

The actual measured value for the acceleration of objects (e.g., apples) dropping near Earth's surface is 9.8 m/sec^2. Thus the relative values of the acceleration of the Moon and of an apple agree with Newton's prediction. It must have been very exciting for Newton when he did this calculation, as it demonstrated that the falling apple and the orbiting Moon could both be explained as due to Earth's gravity.

26. Kepler at Jupiter

26.a Europa's orbit around Jupiter (which is very close to a perfect circle) has a radius of 6.709×10^5 km $= 6.709 \times 10^8$ m. Let us convert this to AU, dividing by 1.496×10^8 km/AU to get 4.485×10^{-3} AU.

Its period is 3.551 Earth days (each day being 86,400 seconds), or 3.068×10^5 seconds, or, using 1 year $\approx 3.156 \times 10^7$ seconds, this is 0.00972 years.

OK, let's now calculate C, using the two sets of units:

$$C = \frac{P^2}{a^3} = 3.117 \times 10^{-16}\,\frac{\mathrm{sec^2}}{\mathrm{m^3}} = 1.047 \times 10^3\,\frac{\mathrm{years^2}}{\mathrm{AU^3}}.$$

Note that we are doing all calculations to four significant figures.

26.b All we need to do is to tabulate the relevant numbers for the other three moons. Here is a table containing the calculations:

Moon	Orbital radius (m)	Period (days)	Period (sec)	P^2/a^3 (sec^2/m^3)
Io	4.216×10^8	1.769	1.528×10^5	3.116×10^{-16}
Ganymede	1.070×10^9	7.155	6.182×10^5	3.120×10^{-16}
Callisto	1.883×10^9	16.69	1.442×10^6	3.114×10^{-16}

So all four moons of Jupiter give the same result, just as Kepler said they should! (Notice that the numbers differ slightly in the third decimal place; this is due to the imprecision of the input numbers, even though we kept four significant figures.)

26.c For units of AU and years, this is a very simple calculation! For Earth going around the Sun, the period is 1 year, and the radius of the orbit is 1 AU, so we get $C = 1^2/1^3 = 1\,\mathrm{year^2/AU^3}$.

26.d We can calculate the ratio of the Cs using either set of units; we will get the same answer (as we must; the ratio has no units). Let's do it in the years-AU units, as the arithmetic will be easy:

$$\frac{C_{\mathrm{Jupiter}}}{C_{\mathrm{Sun}}} = \frac{1.047 \times 10^3\,\mathrm{years^2/AU^3}}{1\,\mathrm{years^2/AU^3}} = 1,047.$$

We can also calculate the ratio of the Sun's mass to that of Jupiter:

$$\frac{M_{\mathrm{Sun}}}{M_{\mathrm{Jupiter}}} = \frac{1.989 \times 10^{30}\,\mathrm{kg}}{1.898 \times 10^{27}\,\mathrm{kg}} = 1,047.$$

It is the same value! Why is this? We can see what's going on from Newton's version of Kepler's third law:

$$P^2 = \frac{4\pi^2 a^3}{GM}.$$

Here we see that what we have called the constant C is in fact $4\pi^2/(GM)$, where M is the mass of the central object about which we are orbiting. This central mass is of course different for the case of the moons of Jupiter, and the orbit of Earth around the Sun. The value of C is inversely proportional to the mass of this central body, and thus the ratio of C for the orbits around Jupiter and the Sun should be the same as the ratio of the mass of the Sun to that of Jupiter, just as we've found.

27. Neptune and Pluto

27.a This is a simple application of Kepler's third law, $P(\text{years}) = a^{3/2}(\text{AU})$. For $a = 30.066$ AU, this gives (yes, we used a calculator) 164.86 years. We give the answer to five significant figures, the same number as given for the semi-major axis.

27.b The same calculation as done in part **a** for Pluto gives 248.1 years, to four significant figures.

27.c The ratio of orbital periods is $248.1/164.86 = 1.505$ to four significant figures. This is quite close (within 0.3%) to a ratio of 3:2. That is, every time Pluto makes two orbits around the Sun, Neptune makes three orbits.

These resonances are actually quite common in the solar system. It turns out that many of the gaps in Saturn's rings are due to resonances with the various moons of Saturn, and more complicated resonances explain some of the stunning detailed features seen in those rings. See, for example, the pictures taken by the Cassini spacecraft at http://saturn.jpl.nasa.gov.

27.d The eccentricity of the orbit is very close to zero, so we're not surprised that the aphelion distance, $a(1+e) = 30.4$ AU, is very close to the value of the semi-major axis. Here, the number of significant figures is subtle. You might think that the eccentricity is known to only a single significant figure, so that the aphelion should be given to the same significance. In fact, what counts in this calculation is the quantity $1 + e$, which has three significant figures.

27.e The perihelion is $a(1-e) = 39.48 \times (1-0.250)\,\text{AU} = 29.61$ AU, and the aphelion is $a(1+e) = 49.35$ AU. The perihelion distance of Pluto is less than the aphelion distance of Neptune, so indeed, Pluto is sometimes a bit closer to the Sun than is Neptune. It only gets a little inside Neptune's orbit, and it was last inside Neptune's orbit from 1979 to 1999.

28. Is there an asteroid with our name on it?

28.a Kepler's third law gives us the semi-major axis:

$$a(\text{AU}) = P^{2/3}(\text{years}).$$

Plugging in numbers (with a calculator), and keeping three significant figures gives

$$a = 1.65\,\text{AU}.$$

But the orbit is not circular. The perihelion is given by $a(1-e)$, and the aphelion by $a(1+e)$. With $e = 0.5$, these multiplicative factors are 0.5 and 1.5, respectively, so we get a perihelion of 0.83 AU and an aphelion of 2.5 AU. So yes, the asteroid's orbit crosses that of Earth; it is an Earth-crosser.

Asteroids don't all orbit in exactly the same plane as does Earth. And to impact Earth, they would have to arrive at 1 AU at exactly the same time Earth reaches that point. So it would seem that it is very unlikely for even an asteroid like 2002 CY9, whose orbit crosses 1 AU, to actually impact Earth. However, an asteroid orbit can be perturbed by the gravitational pull of other bodies, including Earth itself; this will cause its orbit to change slightly over time (tens of years), increasing the chances that it can impact Earth. Calculating these additional perturbations is difficult, with uncertainties that increase with time, so astronomers don't trust their calculations of detailed orbits of Earth-crossers more than about a century into the future.

28.b If we speed up the asteroid by 2 cm/sec, the asteroid will arrive at the point where it would collide with Earth 20 years hence, somewhat sooner than it would have on its original trajectory. If it gets there soon enough, Earth will not have arrived at that spot yet, and thus the asteroid will miss Earth altogether. We will simply ask, what distance does $v = 2$ cm/sec correspond to over $t = 20$ years? If it is larger than Earth's radius, we will be saved. Let us work to a single significant figure:

$$d = vt = 2\,\mathrm{cm/sec} \times 20\,\mathrm{yr} \times \frac{3 \times 10^7\,\mathrm{sec}}{1\,\mathrm{yr}} \times \frac{1\,\mathrm{km}}{10^5\,\mathrm{cm}} \approx 10{,}000\,\mathrm{km}.$$

Earth's radius is 6,400 km, so if the initial trajectory of the asteroid was straight toward Earth, the asteroid would just miss us, if we gave it that nudge. But it would be awfully close; if we were in charge of planning this mission, we'd ask for a more substantial velocity change than that 2 cm/sec.

It is impressive that we could move an asteroid 10,000 kilometers relative to its unperturbed orbit over a 20-year period. If we were in fact threatened by such an impact, we would do a more sophisticated calculation, asking by how much the ellipse around the Sun traced by the asteroid's orbit would change due to the impact, but the calculation you've just done captures the essence of the problem.

29. Halley's comet and the limits of Kepler's third law

29.a Kepler's third law for orbits around the Sun tells us that the period in years is $17.8^{3/2} = 75.1$ years. This is close to, but not identical to, the observed period of 76.1 years. Equivalently, if we start from the period, Kepler's third law gives a semi-major axis of $75.1^{2/3} = 18.0$ AU, again, close but not the same as the observed value of 17.8 AU. Yet another way to do this is to calculate $P^2 = 5{,}790\,\mathrm{yr}^2$, and $a^3 = 5{,}640\,\mathrm{AU}^3$; these two numbers are not the same.

29.b Kepler's third law is a direct consequence of Newton's laws of motion and gravity, where the only relevant gravitational force is due to the Sun. Halley's comet is subject to additional forces from two sources. As Halley comes close to the Sun, jets of gas and dust are ejected. Newton's third law tells us that for every action, there is a reaction; the comet feels a force in reaction to the ejection of those jets. Halley's orbit is additionally perturbed by the gravity from the planets. While the perturbations from the planets can be predicted in detail, the reaction forces due to outgassing from the cometary nucleus are much more difficult, and make it challenging to predict the details of the orbit as it gets close to the Sun.

30. You cannot touch without being touched

30.a The period of the orbit can be calculated directly from Kepler's third law. In particular, we're talking about orbits around the Sun, so we can use the third law in its simplest form:

$$P^2 \text{ (in years)} = a^3 \text{ (in AU)}.$$

We are told that $a = 5.2$ AU; the cube of 5.2 is 140 (we used a calculator), and the square root of that is about 12. Thus the period is 12 years.

30.b Jupiter is moving in a circle of radius 5.2 AU, and it takes 12 years to go all the way around. The speed is straightforward to calculate:

$$v = \frac{2\pi r}{P} \approx \frac{2\pi \times 5.2 \times 1.5 \times 10^{11} \text{ m}}{12 \text{ yr} \times \pi \times 10^7 \text{ sec/yr}} = 13{,}000 \text{ m/sec}.$$

This is a bit less than half the speed of Earth going around the Sun. This is not surprising; Jupiter is farther away from the Sun, thus feels a weaker gravitational pull, and so goes slower.

Here's another way to do it. We know that Earth goes around the Sun at a speed of 30 km/sec. Moreover, we know that the speed is the distance traveled ($2\pi r$) divided by the period P. But Kepler tells us that the period is proportional to $r^{3/2}$, so the speed is proportional to $\frac{r}{r^{3/2}} = r^{-1/2}$. Thus the speed of Jupiter's orbit is slower than that of Earth's orbit by a factor of $5.2^{-1/2} = 0.44$, or 13,000 km/sec.

30.c If the center of mass of the Sun-Jupiter system stays fixed, then when Jupiter moves a certain distance in a certain amount of time, the Sun must move in the opposite sense, so that the quantities $M_{\text{Sun}} \times \text{Motion}_{\text{Sun}}$ and $M_{\text{Jupiter}} \times \text{Motion}_{\text{Jupiter}}$ are equal and opposite. We are interested first in how far the Sun moves:

$$\text{Motion}_{\text{Sun}} = \text{Motion}_{\text{Jupiter}} \times \frac{M_{\text{Jupiter}}}{M_{\text{Sun}}}.$$

Over half its orbit, Jupiter moves the full diameter of its orbit, namely, 10 AU. The ratio of the mass of Jupiter to that of the Sun is 10^{-3}, as stated in the problem. Thus the motion of the Sun is $10^{-3} \times 10 \text{ AU} = 10^{-2} \text{ AU} = 1.5 \times 10^6 \text{ km}$. This turns out to be roughly twice the Sun's radius. So this is a nontrivial motion! Indeed, the Sun carries out this motion along an arc; as Jupiter moves in a circle of diameter 10 AU in its orbit, the Sun moves in a circle of diameter 1.5 million kilometers, or 0.01 AU.

Because each of these motions is happening in the same interval of time, it is also true that $M_{\text{Sun}} \times \text{Speed}_{\text{Sun}}$ and $M_{\text{Jupiter}} \times \text{Speed}_{\text{Jupiter}}$ are equal and opposite. By exactly the same reasoning as above, then, the Sun is moving at 10^{-3} the speed of Jupiter, or about 13 m/sec, a bit faster than the fastest human sprinters.

In fact, astronomers can measure motions of stars as small as 1 m/sec via the Doppler shift and have used this to infer the presence of planets around other stars using exactly the reasoning we've employed here.

31. Aristotle and Copernicus

The Aristotelian worldview had several characteristics:

- Earth lies at the center of the universe.
- Earthbound objects are made of four basic elements: earth, water, air, and fire.

- The natural motion of the elements is toward their "proper place": for earth and water, it is toward the center of Earth, while the natural motion of air and fire is away from the center of Earth.
- The material of the heavens, or celestial spheres, is made of a fifth element, different from that found on Earth, namely, quintessence.
- The natural motion of celestial objects is in circles; indeed, the celestial spheres rotate around Earth.
- The heavens are unchanging.

Indeed for our purposes, the most important aspect of the Aristotelian worldview is that the substance that objects in the heavens were made of, and the physical laws that governed those objects, were different from those on Earth.

Perhaps the most profoundly revolutionary aspect of the Scientific Revolution, as exemplified by the work of Kepler, Copernicus, Newton, Galileo, and Payne-Gaposchkin, was the understanding that the substance of heavenly bodies and the laws that govern their motions are the same as those on Earth. Let's take the work of each of these great scientists in turn:

- Nicolas Copernicus showed that motions in the solar system are much more easily explained if instead of placing Earth at the center, the Sun lies at the center of the solar system. This contradicts one of the Aristotelian notions (i.e., that all heavenly objects move in circles around Earth).
- Galileo Galilei was the first to use a telescope to study the sky. His discoveries were in strong support of Copernicus' ideas and served to further undermine the Aristotelian notions of the heavenly objects as perfect:

  - He saw mountains on the Moon, very much like those on Earth.
  - He saw spots on the Sun, demonstrating that it was not unblemished.
  - He saw that Venus went through phases, easily explained (and predicted!) in the heliocentric view, and not expected in the geocentric view.
  - He saw that Jupiter had a set of four moons orbiting around it, showing directly that heavenly bodies can exhibit circular motions around objects other than Earth.

- Johannes Kepler developed three laws of planetary motion, which he determined empirically from detailed observations made by Tycho Brahe. They were:

  - Planets move in elliptical orbits, with the Sun at one focus.
  - A planet sweeps out equal areas in equal time.
  - The square of the period of a planet's orbit (in years) is equal to the cube of its semi-major axis (in AU).

- Isaac Newton developed his laws of motion, and his laws of gravity, showing explicitly that the laws that govern motions on Earth and in the heavens are the same. In particular, he showed that the acceleration that pulls an apple to the ground is due to the same force of gravity that keeps the Moon "falling" in its orbit around Earth.
- Finally, Cecilia Payne-Gaposchkin studied the spectra of stars and used the (then new-fangled) notions of quantum mechanics to infer the chemical composition of stars. The important point here was that stars are made up of the same elements that are found on Earth, going against one of the basic Aristotelian tenets. In

particular, the Sun, and stars in general, are composed of 74% hydrogen and 24% helium by mass, leaving 2% for all the rest of the elements in the periodic table.

32. Distant supernovae

The relationship between brightness b, luminosity L, and distance d is

$$b = \frac{L}{4\pi d^2}.$$

We write this relationship down for α Centauri and the 1987 supernova, and take the ratio of the two expressions to find

$$\frac{b_{\text{supernova}}}{b_{\alpha\,\text{Cen}}} = \frac{L_{\text{supernova}}}{L_{\alpha\,\text{Cen}}} \times \left(\frac{d_{\text{supernova}}}{d_{\alpha\,\text{Cen}}}\right)^{-2}.$$

The ratio of brightnesses is $1/10$, and the ratio of distances is $150,000/4 \approx 4 \times 10^4$, to one significant figure. Solving for the ratio of the luminosities, we get

$$\frac{L_{\text{supernova}}}{L_{\alpha\,\text{Cen}}} = (4 \times 10^4)^2 \times \frac{1}{10} = 1.6 \times 10^8.$$

The supernova peaked at a luminosity of an astonishing 160 million times that of α Cen.

33. Spacecraft solar power

33.a This is a straightforward application of the blackbody formula, which states that the luminosity of a spherical blackbody of temperature T and radius R is

$$L = 4\pi R^2 \sigma T^4.$$

Here, $\sigma = 5.6 \times 10^{-8}$ Joules/sec/m^2/K^4 is the Stefan-Boltzmann constant. So solving this problem is just a matter of plugging in the numbers, which we will try without a calculator:

$$L = 4 \times 3 \times (7 \times 10^8\,\text{m})^2 \times 5.6 \times 10^{-8}\,\text{Joules/sec/m}^2/\text{K}^4 \times (6{,}000\,\text{K})^4 \approx$$

$$\approx 12 \times 50 \times 5.6 \times 1{,}300 \times 10^{16-8+12}\,\text{Joules/sec} \approx 4 \times 10^{26}\,\text{watts}$$

to a single significant figure. This is very close to the true value of 3.8×10^{26} watts.

33.b This is an exercise in the inverse square law. The brightness is simply given by the luminosity, divided by $4\pi d^2$, where $d = 1.5$ AU is the distance of Mars from the Sun:

$$b = \frac{L}{4\pi d^2} = \frac{4 \times 10^{26}\,\text{watts}}{12 \times (1.5 \times 1.5 \times 10^{11}\,\text{m})^2} = 660\,\text{watts/m}^2.$$

33.c The above-calculated power is incident on every square meter of the solar panels. We first need to calculate the number of square meters 100 square feet are. 1 foot $= 1/3$ meter, so 1 foot^2 $= 1/9$ meter^2. So 100 square feet $= 10$ meters^2, to one significant figure. Thus the total power hitting the solar arrays is 10 times the value we calculated above, or 6,600 watts. However, only 20%, or 1/5, of that is available to the transmitter, namely, 1,300 watts.

33.d The inverse square law says that the power available to the same spacecraft at a larger distance is reduced by the inverse square of the ratio of the distances. That is, if we went through the calculation we did in parts **b** and **c**, and used the distance to Saturn rather than Mars, everything would be identical but the distance itself, and the results would be proportional to the inverse square of the distance. So we would find that the power available to the spacecraft at Saturn would be

$$\text{Power at Saturn} = \text{Power at Mars} \times \left(\frac{\text{Distance to Mars}}{\text{Distance to Saturn}}\right)^2 = 1,300\,\text{watts}\left(\frac{1.5\,\text{AU}}{9.6\,\text{AU}}\right)^2.$$

The ratio of distances is roughly 1/6.5, whose square is about 1/40. So the available power at Saturn's distance is about 30 watts, or 1/3 of an ordinary light bulb. Scarcely enough to run a spacecraft! For this reason, missions to the outer solar system often rely on power generated from the radioactive decay of a plutonium isotope, rather than from solar panels.

34. You glow!

The luminosity of a blackbody is

$$F = \sigma T^4 \times A,$$

where σ is the Stefan-Boltzmann constant, and A is the surface area of the object. First, let's estimate a person's temperature. We know that water freezes at 273 K, and that human body temperature is about 37 K higher than that. We just want a crude estimate, so $T = 300$ K is close enough. To estimate surface area, we consider that a person is about a meter tall and about a meter wide. Of course, most people are more than a meter tall and less than a meter wide, but again, we just want an estimate, and these values are certainly accurate to a factor of about 2. Let's also multiply by 2, because people have a front and a back side, and they will radiate from both. So

$$T = 300 \text{ K},$$

$$\text{Surface area} = 2 \text{ m}^2.$$

Hence,

$$\text{Luminosity} = \sigma T^4 \times \text{Area}$$

$$= 6 \times 10^{-8} \text{ W m}^{-2} \text{ K}^{-4} (300 \text{ K})^4 \times 2 \text{ m}^2$$

$$\approx 900 \text{ W},$$

or the equivalent of about nine ordinary light bulbs.

With a surface temperature of 300 K, we radiate our blackbody energy in the infrared part of the spectrum. Our eyes are not sensitive to infrared light, so we appear invisible in a darkened room. So-called heat-sensitive goggles, which convert infrared light into visible light, do allow us to see the blackbody radiation from living creatures, even in complete darkness.

35. Tiny angles

35.a We are asked how long it takes a star moving at a certain speed to traverse a certain angle. If we can convert that angle into a distance, then we're home-free: time = distance/speed. So let's start by calculating that distance.

If a star at a distance of 300 light-years appears to have moved 1 arcsecond, it has moved $1/(2 \times 10^5)$ radian, corresponding to a physical distance (according to the small-angle formula) of

$$s = \theta \times d = \frac{300\,\mathrm{ly} \times 10^{13}\,\mathrm{km/ly}}{2 \times 10^5} = 1.5 \times 10^{10}\,\mathrm{km}.$$

Here we've used the conversion between kilometers and light-years. Now, it's moving at 300 kilometers per second, and the question asks how far it will take to move that distance:

$$\mathrm{Time} = \frac{\mathrm{Distance}}{\mathrm{Speed}} = \frac{1.5 \times 10^{10}\,\mathrm{km}}{300\,\mathrm{km/sec}} = 5 \times 10^7\,\mathrm{sec} \approx 1.5\,\mathrm{years}.$$

Most nearby stars (i.e., closer than 300 light-years) are part of the Milky Way disk and thus are in similar orbits to the Sun around the center of the Milky Way. So their motion relative to us is relatively slow, tens of kilometers per second. In contrast, those stars moving as fast as 300 kilometers per second are members of the halo of the Milky Way. Their stellar orbits are more randomly oriented, and thus the difference in velocity of such a star relative to the Sun can be quite large.

35.b In 1 year, the star moves 10 arcseconds at a distance of 6 light-years, or about 2 parsecs (we're working here to one significant figure). We know that 1 AU subtends 1 arcsecond at 1 parsec, so 10 arcseconds, at 2 parsecs, subtends 20 AU. That is, the star moves 20 AU in a year. Thus, the speed of the star is

$$v = \frac{d}{t} = \frac{20\,\mathrm{AU} \times 1.5 \times 10^8\,\frac{\mathrm{km}}{\mathrm{AU}}}{1\,\mathrm{year} \times 3 \times 10^7\,\frac{\mathrm{sec}}{\mathrm{year}}} = 100\,\frac{\mathrm{km}}{\mathrm{sec}}.$$

Arithmetic to a single significant figure is easy!

There is a subtlety here. The apparent motion in the sky (what is often referred to as *proper motion*) reflects only that component of the motion of the star in the plane of the sky. The motion along the line of sight, perpendicular to the plane of the sky, is not measurable this way, but it does manifest itself by the Doppler effect, measurable from observations of the star's spectrum. So the combination of the measurement of the spectrum and the proper motion of the star, and the star's distance, allow one to infer its three-dimensional motion through space.

36. Thinking about parallax

36.a If the star lies in the ecliptic, then it will only appear to move back and forth in a straight line, also along the ecliptic (see figure 9.a). The definition of 1 parsec is that a star at this distance moves through an angular distance of 1 arcsecond as the observer moves a distance of 1 AU. Since Earth's orbit moves over a maximum distance of 2 AU (from one side of the sun to the other), a star at a distance of 1 parsec will move through an angular distance of 2 arcseconds over the course of the year. Hence, a star at a distance of 10 parsecs will move through an angular distance of 0.2 arcseconds.

36.b If the star lies at the ecliptic pole, then it will appear to move in a circle relative to the background stars (figure 9.b). The diameter of the circle will also be 0.2 arcseconds.

36.c If a star has a proper motion, then that motion will simply be added to the motion due to parallax. For instance, if the proper motion is perpendicular to the ecliptic,

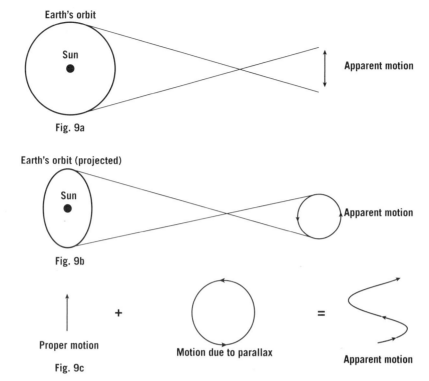

Figure 9. Figure for problem 36.

then the star would appear to move in a sinusoidal motion across the sky (see figure 9.c).

37. Really small angles and distant stars

37.a Consider a long skinny triangle, with acute angle θ equal to 10 micro-arcseconds and length $d = 4,000$ kilometers. We have to convert the angle to radians, of course. The small-angle formula tells us that the far side has a length:

$$s = d\theta = 4 \times 10^3 \, \text{km} \times 10^3 \, \frac{\text{m}}{\text{km}} \times 10^6 \, \frac{\text{micron}}{\text{m}} \times 10^{-5} \, \text{arcsec} \times \frac{1 \, \text{radian}}{2 \times 10^5 \, \text{arcsec}} = 200 \, \text{microns}.$$

200 microns is the size of an amoeba. That is, 10 micro-arcseconds is the angle an amoeba in Los Angeles would subtend as seen from New York.

37.b The small-angle formula $\theta = s/d$ tells us that the parallax θ is inversely proportional to its distance d. The definition of a parsec tells us that a star with a parallax of 1 arcsecond lies at a distance of 1 parsec. So the distance of a star with a 50 micro-arcsecond $= 5 \times 10^{-5}$ arcsec parallax is simply

$$\frac{1}{5 \times 10^{-5}} \, \text{parsec} = 20,000 \, \text{parsecs},$$

or 20 kiloparsecs, which is comparable to the full extent of the Milky Way. This is one of the reasons astronomers are so excited about Gaia; for the first time, we'll get measurements of distances for stars over an appreciable fraction of the Milky Way Galaxy.

38. Brightness, distance, and luminosity

38.a Let's start by discussing what's physically going on here. We're told the star's radius R and surface temperature T, and we know a formula for the luminosity of a spherical blackbody given those two quantities:

$$L = 4\pi R^2 \sigma T^4.$$

So we know how to calculate the luminosity. We have a measured brightness b as well, and the inverse square law relates the luminosity and distance to the brightness, so we'll be able to determine the distance.

For this problem, you do not need to know the numerical value of σ (the Stefan-Boltzmann constant), which appears in the blackbody formula. Let's think about that, by writing the blackbody formula down twice, once for the star ($*$) and once for the Sun:

$$L_* = 4\pi R_*^2 \sigma T_*^4.$$

$$L_{\text{Sun}} = 4\pi R_{\text{Sun}}^2 \sigma T_{\text{Sun}}^4.$$

We're now going to take the ratio of these two formulas; as we'll see, this makes our problem much simpler computationally:

$$\frac{L_*}{L_{\text{Sun}}} = \left(\frac{R_*}{R_{\text{Sun}}}\right)^2 \left(\frac{T_*}{T_{\text{Sun}}}\right)^4.$$

The σ dropped out! Moreover, the ratios of radii and temperatures are easy, so the ratio of luminosities is:

$$\frac{L_*}{L_{\text{Sun}}} = \left(\frac{1}{10}\right)^2 \left(\frac{1}{2}\right)^4 = 10^{-2} \times \frac{1}{16} = 6 \times 10^{-4}.$$

This star is near the bottom of the main sequence; such stars are much less luminous than is the Sun. The luminosity is

$$L_* = 6 \times 10^{-4} L_{\text{Sun}} = 6 \times 10^{-4} \times 4 \times 10^{33} \, \text{erg/sec} = 2.4 \times 10^{30} \, \text{erg/sec}.$$

What is the distance d of such a star? The inverse square law states

$$b = \frac{L}{4\pi d^2},$$

or

$$d = \left(\frac{L}{4\pi b}\right)^{1/2}.$$

Let's plug in numbers:

$$d = \left(\frac{2.4 \times 10^{30} \, \text{erg/sec}}{4\pi \times 6 \times 10^{-11} \, \text{erg/sec/cm}^2}\right)^{1/2}.$$

$6 \times 4 = 24$, and $10/\pi \approx 3$, so the arithmetic isn't too bad:

$$d = \left(3 \times 10^{29-1+11} \, \text{cm}^2\right)^{1/2} = \left(30 \times 10^{38} \, \text{cm}^2\right)^{1/2} \approx 5 \times 10^{19} \, \text{cm}.$$

Remembering that there are about 10^{18} centimeters in a light-year, the distance is about 50 light-years. This is pretty nearby, as stars go.

We also know there are 3.26 light-years in a parsec, so our calculated d is about 15 parsecs. The parallax in arcseconds is the inverse of the distance in parsecs, so this star has a parallax of about 0.06 arcseconds, or 60 milli-arcseconds.

38.b The distance depends on the luminosity and the brightness. What is the luminosity of the star? By exactly the same argument as in part **a**, its luminosity relative to the Sun is

$$\frac{L_*}{L_{\text{Sun}}} = \left(\frac{R_*}{R_{\text{Sun}}}\right)^2 \left(\frac{T_*}{T_{\text{Sun}}}\right)^4 = (200)^2 \times \left(\frac{4,500}{5,800}\right)^4 \approx 14,000.$$

We've used the fact that 1 AU is about 200 times the radius of the Sun (and here we did use a calculator). We could calculate its luminosity and thus its distance as done before, but we're going to proceed in a slightly different way: let's do a bit more algebraic work to avoid unnecessary arithmetic. The spirit of these tricks is to look at the work already done and try to use it to the extent possible to not repeat calculations unnecessarily. Remember that these two stars have the same brightness. So we can write down the relationship between the distance, luminosity, and brightness of each star:

$$d_{\text{star 1}} = \left(\frac{L_{\text{star 1}}}{4\pi b}\right)^{1/2}$$

$$d_{\text{star 2}} = \left(\frac{L_{\text{star 2}}}{4\pi b}\right)^{1/2}$$

where "star 1" refers to the cool main sequence star of part **a**, and "star 2" is the red giant. The brightnesses b of the two are the same. Again, take the ratios of these two expressions to find:

$$\frac{d_{\text{star 2}}}{d_{\text{star 1}}} = \left(\frac{L_{\text{star 2}}}{L_{\text{star 1}}}\right)^{1/2}.$$

Plugging in the numbers, where we'll use luminosities relative to the Sun, we get:

$$\frac{d_{\text{star 2}}}{d_{\text{star 1}}} = \left(\frac{1.4 \times 10^4}{6 \times 10^{-4}}\right)^{1/2} \approx (2 \times 10^7)^{1/2} \approx 4,000.$$

Thus these two stars, of the same brightness, have distances differing by a factor of 4,000! Star 1 is 50 light-years away, so star 2 is 4000 times farther away, 200,000 light-years away, or about 60 kiloparsecs. Such a star must be in the outer halo of the Milky Way. Astronomers are very interested to know the distribution of stars in the outer Milky Way. Finding such stars isn't easy: most red stars in the sky of this brightness are nearby main sequence stars rather than very distant red giants. Even though the parallax of such a distant red giant star will be unmeasurable even by the Gaia satellite, the most sensitive telescope ever devised for this purpose (see problem 37), the fact that the parallax is very small will tell us that the star must be very distant, more than 20 kiloparsecs away. Thus, given its measured brightness, its inferred luminosity must be very large; it cannot be a main sequence star and so must be a red giant.

39. Comparing stars

39.a The relation between peak wavelength and surface temperature is

$$\lambda_{\text{peak}} = \frac{2.9\,\text{mm}}{T}.$$

For a temperature of 3×10^4 K, this gives 1×10^{-4} millimeters to one significant figure. There are 1,000 microns per millimeter, so the wavelength is 0.1 microns. This is 1,000 Ångstroms, which is in the ultraviolet part of the spectrum.

39.b Here we recognize that the luminosities of these stars are given by the blackbody formula:

$$L_{\text{O star}} = 4\pi R_{\text{O star}}^2 \sigma T_{\text{O star}}^4;$$

$$L_{\text{red giant}} = 4\pi R_{\text{red giant}}^2 \sigma T_{\text{red giant}}^4.$$

We can get a handle on the ratio of the radii of the two stars by taking the ratio of the two equations. When we do so, lots of terms drop out, and we're left with

$$\frac{L_{\text{red giant}}}{L_{\text{O star}}} = \left(\frac{R_{\text{red giant}}}{R_{\text{O star}}}\right)^2 \left(\frac{T_{\text{red giant}}}{T_{\text{O star}}}\right)^4.$$

Solving for the ratio of radii gives

$$\frac{R_{\text{red giant}}}{R_{\text{O star}}} = \left(\frac{L_{\text{red giant}}}{L_{\text{O star}}}\right)^{1/2} \left(\frac{T_{\text{O star}}}{T_{\text{red giant}}}\right)^2.$$

We know the ratio of the luminosities, namely, 16. And the ratio of the temperatures is $30,000/3,000 = 10$. So we find

$$\frac{R_{\text{red giant}}}{R_{\text{O star}}} = (16)^{1/2} \times 10^2 = 400.$$

39.c The brightnesses of the two stars are related to their distances (which is we want to know) and their luminosities (which we know something about):

$$b_{\text{red giant}} = \frac{L_{\text{red giant}}}{4\pi d_{\text{red giant}}^2}$$

$$b_{\text{O star}} = \frac{L_{\text{O star}}}{4\pi d_{\text{O star}}^2}$$

But we are told that $b_{\text{red giant}} = b_{\text{O star}}$. So we can equate these two expressions and solve for the distance to the red giant:

$$\frac{L_{\text{red giant}}}{4\pi d_{\text{red giant}}^2} = \frac{L_{\text{O star}}}{4\pi d_{\text{O star}}^2}.$$

A little algebra gives

$$d_{\text{red giant}} = d_{\text{O star}} \left(\frac{L_{\text{red giant}}}{L_{\text{O star}}}\right)^{1/2} = 1,000\,\text{ly} \times \sqrt{16} = 4,000\,\text{ly}.$$

40. Hot and radiant

40.a The luminosity of a blackbody of surface temperature T and radius R is proportional to $R^2 T^4$, by the Stefan-Boltzmann law. So the ratio of the luminosities is

$$\frac{L_{\text{Sun}}}{L_{\text{Jupiter}}} = \left(\frac{R_{\text{Sun}}}{R_{\text{Jupiter}}}\right)^2 \left(\frac{T_{\text{Sun}}}{T_{\text{Jupiter}}}\right)^4.$$

Note that we first take the ratio of radii, then square. Similarly, we take the ratio of temperatures before raising to the fourth power. This will make the arithmetic much easier. Plugging in numbers, we get

$$\frac{L_{\text{Sun}}}{L_{\text{Jupiter}}} = (10)^2 \left(\frac{6,000\,\text{K}}{200\,\text{K}}\right)^4 = 100 \times 30^4 = 8 \times 10^7,$$

to one significant figure. Wow, the Sun is almost 100 million times more luminous than is Jupiter!

Wein's law tells us the wavelength at which the emission from Jupiter peaks:

$$\lambda_{\text{peak}} = \frac{0.29\,\text{cm}}{T} = \frac{0.29\,\text{cm}}{200} = 1.5 \times 10^{-3}\,\text{cm} = 15\,\text{microns}.$$

This is light in the infrared part of the spectrum.

Our eyes are insensitive to infrared light, but Jupiter is easy to see with the naked eye at night. What we're seeing is not the light emitted by the planet, but rather the light reflected by it from the Sun. It turns out that the total luminosity of reflection and of blackbody emission for Jupiter are roughly comparable.

40.b This is a similar problem to that in part **a**, and we'll do it in the same way. The radius is half the diameter, namely, 7,000 km. The ratio of luminosities is

$$\frac{L_{\text{WD}}}{L_{\text{Sun}}} = \left(\frac{R_{\text{WD}}}{R_{\text{Sun}}}\right)^2 \left(\frac{T_{\text{WD}}}{T_{\text{Sun}}}\right)^4 \approx \left(\frac{7,000\,\text{km}}{700,000\,\text{km}}\right)^2 \left(\frac{40,000\,\text{K}}{6,000\,\text{K}}\right)^4 = \left(\frac{1}{100}\right)^2 (7)^4 \approx \frac{1}{4}.$$

Even though the white dwarf is much smaller than the Sun, it is only a factor of 4 less luminous, because of its much higher temperature. The peak wavelength is

$$\lambda_{\text{peak}} = \frac{0.29\,\text{cm}}{T} = \frac{0.29\,\text{cm}}{40,000} = 7.5 \times 10^{-6}\,\text{cm} = 0.075\,\text{microns}.$$

This is in the ultraviolet part of the spectrum. Even though our eyes are not sensitive to ultraviolet light, the star will still put out plenty of visible light, more blue than red, so the star will appear blue.

40.c This is very similar to the previous calculation. We write

$$\frac{L_{\text{red giant}}}{L_{\text{Sun}}} = \left(\frac{R_{\text{red giant}}}{R_{\text{Sun}}}\right)^2 \left(\frac{T_{\text{red giant}}}{T_{\text{Sun}}}\right)^4.$$

We are told that the ratio of the diameters (and therefore the radii) is 80. The Sun has a surface temperature of 6,000 K, and therefore the ratio of temperatures is 1/2. Thus we have

$$\frac{L_{\text{red giant}}}{L_{\text{Sun}}} = (80)^2 \times \left(\frac{1}{2}\right)^4.$$

$2^4 = 16$, so this becomes

$$\frac{L_{\text{red giant}}}{L_{\text{Sun}}} = \frac{8 \times 8 \times 100}{16} = 400,$$

without having to touch a calculator. By Wein's law, the peak wavelength is

$$\lambda_{\text{peak}} = \frac{0.29\,\text{cm}}{3000} = 1\,\text{micron}.$$

This wavelength is in the near-infrared part of the spectrum. Our eyes are not sensitive to light of wavelength 1 micron, but remember that this is the peak of the spectrum. That is, the spectrum rises from blue wavelengths to red into infrared wavelengths. There is substantially more light coming out at red wavelengths than at blue wavelengths, so the object appears red to our eyes (which is why we call it a red giant!).

41. A white dwarf star

41.a The inverse square law relates brightness b, luminosity L, and distance r:

$$b = \frac{L}{4\pi r^2}.$$

We can write such an equation down separately for the Sun and for the white dwarf. If we take the ratio of the two, realizing that the luminosities of the two are the same, we find:

$$\left(\frac{r_{\text{WD}}}{r_{\text{Sun}}}\right)^2 = \frac{b_{\text{Sun}}}{b_{\text{WD}}} = 10^{16}.$$

Taking the square root of both sides, and solving for r_{WD} yields

$$r_{\text{WD}} = 10^8 r_{\text{Sun}} = 10^8\,\text{AU}.$$

There are 60,000 AU in a light-year, so we find

$$r_{\text{WD}} = 1,700\,\text{ly}.$$

41.b We know that the Sun and the white dwarf have the same luminosity. We also know that for blackbodies, luminosities are related to radius and temperature:

$$L = 4\pi R^2 \sigma T^4.$$

Writing down such an equation for both the white dwarf and the Sun, and taking their ratio gives

$$1 = \left(\frac{R_{\text{WD}}}{R_{\text{Sun}}}\right)^2 \left(\frac{T_{\text{WD}}}{T_{\text{Sun}}}\right)^4.$$

The temperature ratio we know; the white dwarf is 10 times hotter than the Sun. Taking the square root of both sides, and solving for R_{WD} gives

$$R_{\text{WD}} = R_{\text{Sun}} \times 10^{-2} = 7 \times 10^5\,\text{km}/100 = 7,000\,\text{km}.$$

The white dwarf is about the size of Earth.

42. Orbiting a white dwarf

42.a Newton's form of Kepler's third law states that for orbits around a body of mass M, the orbital period P is related to the radius of the orbit a by

$$P^2 \propto \frac{a^3}{M}.$$

Therefore the ratio of the orbital period around the white dwarf to that of the space shuttle is

$$\frac{P_{WD}^2}{P_{Earth}^2} = \left(\frac{a_{WD}^3}{M_{WD}}\right) \Big/ \left(\frac{a_{Earth}^3}{M_{Earth}}\right) = \frac{M_{Earth}}{M_{WD}},$$

where we used the fact that the radii of the two orbits are the same and thus drop out of the ratio.

Solving for the period of the orbit around the white dwarf, we find

$$P_{WD} = P_{Earth}\sqrt{\frac{M_{Earth}}{M_{WD}}} = 90\,\mathrm{min}\sqrt{\frac{6 \times 10^{24}\,\mathrm{kg}}{2 \times 10^{30}\,\mathrm{kg}}} = 90\,\mathrm{min} \times \sqrt{3 \times 10^{-6}},$$

$$P_{WD} = 90\,\mathrm{min} \times \frac{60\,\mathrm{sec}}{1\,\mathrm{min}} \times 1.7 \times 10^{-3} = 9\,\mathrm{sec},$$

where we used the useful fact that $60 \times 1.7 \approx 100$.

The hard way to do this problem is to use Kepler's third law without making reference to any proportionalities and calculate things out in full glory. For completeness, and to show that this can also be done without a calculator, let's give it a try. Newton's form of Kepler's third law is

$$P_{WD} = 2\pi\sqrt{\frac{R_{WD}^3}{GM_{WD}}}.$$

Plugging in numbers and being careful to keep the units consistent, we get

$$P_{WD} = 6 \times \sqrt{\frac{(6.4 \times 10^6\,\mathrm{m})^3}{(\frac{2}{3} \times 10^{-10}\,\mathrm{m}^3\,\mathrm{sec}^{-2}\,\mathrm{kg}^{-1})(2 \times 10^{30}\,\mathrm{kg})}}$$

$$= 6 \times \sqrt{\frac{240 \times 10^{18}\,\mathrm{m}^3}{1.3 \times 10^{20}\,\mathrm{m}^3\,\mathrm{sec}^{-2}}},$$

$$P_{WD} = 6 \times \sqrt{1.8}\,\mathrm{s} = 9\,\mathrm{sec}.$$

We get the same answer, but the arithmetic (still done without a calculator!) was quite a bit more difficult.

42.b Astronauts in such a spaceship would indeed feel weightless. The astronauts are in orbit around the white dwarf, just as the spaceship is. The astronauts and the spaceship are in free fall together, at the same rate, around the white dwarf (since they are falling toward the white dwarf at the same rate as the white dwarf curves away from them). There is no difference in the accelerations of the spaceship, the astronauts, and everything inside the spaceship. Therefore no net force pushes the astronauts toward the spaceship's walls or floor: they experience weightlessness.

43. Hydrogen absorbs

43.a The first thing to do is to find the wavelengths of the absorption lines. We found the following wavelengths corresponding to the bottoms (i.e., centers) of the absorption lines:

370.5 nanometers (nm), 371.0 nm, 372.1 nm, 373.4 nm, 374.9 nm, 377.0 nm, 379.7 nm, 383.5 nm, 388.8 nm, 393.3 nm, 396.9 nm, 410.1 nm, 434.0 nm, 486.0 nm, 656.3 nm, and 687.1 nm.

The line at 687.1 nm is not due to hydrogen; it is a line due to water vapor in Earth's atmosphere. The next thing to do is to note the hint that the α line (for which $m = n+1$) is included in this list. Staring at the given formula for wavelength λ, you can convince yourself that this is the line that is longest in wavelength, and therefore is the line at 656.3 nm, which must be equal to (after a bit of algebra):

$$\lambda_{\text{longest}} = \frac{1}{R} \frac{n^2(n+1)^2}{(2n+1)}.$$

Just as the longest wavelength corresponds to $m = n+1$ (the smallest possible value for m), the shortest possible wavelength corresponds to a very large value for m. Indeed, if m is infinite, then $\lambda_{\text{shortest}} = n^2/R$. We can estimate $\lambda_{\text{shortest}}$ by noticing that the absorption lines are getting closer and closer as one goes to short wavelengths; there is a limit near $\lambda_{\text{shortest}} \approx 370\,\text{nm}$.

Let us look at the ratio between these two wavelengths (notice that R drops out when we do this which is a good thing, because we don't know R yet). The observed ratio is 1.77. The theoretical ratio is

$$\frac{\lambda_{\text{longest}}}{\lambda_{\text{shortest}}} = \frac{n^2(n+1)^2/(2n+1)/R}{n^2/R}$$

$$= \frac{(n+1)^2}{2n+1}.$$

For $n = 1$, this ratio is $4/3 = 1.33$.
For $n = 2$, this ratio is $9/5 = 1.80$.
For $n = 3$, this ratio is $16/7 = 2.29$.
For $n = 4$, this ratio is $25/9 = 2.78$.
For $n = 5$, this ratio is $36/11 = 3.27$.

The observed ratio is 1.77, a close match to the above for $n = 2$. Now we can solve for R by looking at the α line, for which $n = 2, m = 3$:

$$R = \frac{36/5}{656.3\,\text{nm}}$$

$$= 1.097 \times 10^7\,\text{m}^{-1},$$

where we converted from nanometers to meters. Let us now calculate the expected values of wavelength for a range of values of m:

n	m	λ_{calc} (nm)	λ_{obs} (nm)
2	3	656.3	656.3
2	4	486.2	486.0
2	5	434.1	434.0
2	6	410.2	410.1
2	7	397.0	396.9
2	8	388.9	388.8
2	9	383.6	383.5
2	10	379.8	379.7
2	11	377.1	377.0
2	12	375.0	374.9
2	13	373.5	373.4
2	14	372.2	372.1
2	15	371.2	371.0
2	16	370.4	370.5

The match between calculated and observed wavelengths is essentially perfect! The series continues in principle to $m = \infty$, with the lines becoming more and more crowded together, but this spectrum does not have the resolution to show this. Only two lines in our list do not follow this pattern: 687.1 nm is due to water vapor in Earth's atmosphere and the line at 393.3 nm is due to calcium in the atmosphere of the star (there is in fact another line of calcium that coincides almost perfectly with the $m = 7$ line of hydrogen). You'll notice that both these lines have a different shape from the hydrogen lines, which is one way to distinguish them.

This $n = 2$ series of hydrogen is called the Balmer series, after the scientist who first described it. There are other series, corresponding to other values of n: the Lyman series ($n = 1$) is in the far-ultraviolet; the Paschen series ($n = 3$) is in the near-infrared, and so on.

43.b The spectrum peaks at about 400 nanometers, or 4×10^{-5} cm. The relation between the peak wavelength and the temperature of a star is

$$T \approx \frac{0.3\,\text{cm}}{\lambda_{\text{peak}}} = \frac{0.3\,\text{cm}}{4 \times 10^{-5}\,\text{cm}} \approx 7,500\ \text{K},$$

which is indeed close to the nominal temperature for an F star.

44. The shining Sun

44.a To calculate the energy produced when 1 gram of hydrogen is burned in the Sun into helium, we first calculate the energy released in the creation of one helium atom:

$$\Delta E = m_{\text{4H}}c^2 - m_{\text{He}}c^2 = (6.693 \times 10^{-27}\,\text{kg} - 6.645 \times 10^{-27}\,\text{kg})(3.0 \times 10^8\,\text{m/sec})^2$$
$$= 4.3 \times 10^{-12}\ \text{Joules}.$$

Then the total amount of energy produced if 1 gram of hydrogen is burned into helium is the amount of energy produced per atom times the total number of helium atoms (or equivalently, the number of hydrogen atoms divided by 4) in 1 gram. The latter quantity is simply the reciprocal of the mass of the helium atom (expressed in grams, not kilograms), so we get

$$E_{1\text{ gram H}} = N_{\text{He}}\Delta E = \frac{1\text{ gram}}{m_{\text{He}}}\Delta E = (6.6 \times 10^{-24})^{-1}(4.3 \times 10^{-12}\text{ Joules})$$

$$= 6.4 \times 10^{11}\text{ Joules}.$$

44.b The maximum energy the Sun could produce over its lifetime by burning hydrogen is its mass of hydrogen in grams times the energy released in burning 1 gram of hydrogen:

$$E_{\text{Sun,H}} = \frac{3}{4}M_{\text{Sun}}E_{1\text{ gram H}} = \frac{3}{4}(2 \times 10^{33}\text{ g})(6.4 \times 10^{11}\text{ Joules/g}) \approx 10^{45}\text{ Joules}.$$

(Note that we converted the mass of the Sun to grams.)

We can therefore estimate the maximum time the Sun could shine at its present luminosity by burning hydrogen into helium as

$$t = \frac{E_{\text{Sun,H}}}{L_{\text{Sun}}} = \frac{10^{45}\text{ Joules}}{4 \times 10^{26}\text{ Joules/sec}} = 2.5 \times 10^{18}\text{ sec} \approx 7 \times 10^{10}\text{ years}.$$

The main sequence lifetime of a solar type star is only 10^{10} years. But only the hydrogen in the core of the Sun is hot enough to burn, so the Sun ends its main sequence lifetime when roughly 10% of its hydrogen has burned to helium.

45. Thermonuclear fusion and the Heisenberg uncertainty principle

45.a We are asked to equate the initial kinetic energy $3kT$ (remember that there are two protons!) with the final potential energy $q\frac{e^2}{r}$ and to solve for the temperature. Let's first do this algebraically:

$$3kT = q\frac{e^2}{r},$$

$$T = \frac{qe^2}{3kr},$$

and then plug in numbers. Here we take r as the diameter of a proton (i.e., 10^{-15} meters):

$$T = \frac{1}{3} \times 9 \times 10^9 \times \frac{\left(\frac{1}{6} \times 10^{-18}\right)^2}{1.4 \times 10^{-23} \times 10^{-15}}\text{ Joule meter Coulomb}^{-2} \times \text{Coulomb}^2 \times \frac{\text{K}}{\text{Joule meter}},$$

where we've gathered all the units together at the end of the above expression. To do this arithmetic, pull out all the factors of 10, yielding

$$\frac{1}{3} \times 9 \times \left(\frac{1}{6}\right)^2 \times 10^{9-36+22+15}\text{ K} = \frac{1}{2} \times 10^{10}\text{ K} = 5 \times 10^9\text{ K},$$

or 5 billion K. To simplify the arithmetic, we approximated $6 \times 1.4 \approx 10$.

Five billion K is hot, indeed much too hot; the core of the Sun is about 15 million K, a factor of 300 times smaller than our computed value. So on the face of it, it seems that the Sun is simply not hot enough for thermonuclear fusion to take place.

45.b This problem has several parts. First, we are asked to check the units in the expression above for the de Broglie wavelength. Remembering that 1 Joule $= 1\,\text{kg}\,\text{m}^2/\text{sec}^2$, we find that the quantity $\frac{h}{mv}$ has units:

$$\frac{\text{Joule}\,\text{sec}}{\text{kg}\,\text{m}/\text{sec}} = \frac{\text{kg}\,\text{m}^2}{\text{sec}} \times \frac{\text{sec}}{\text{kg}\,\text{m}} = \text{meters},$$

which is what we expected.

The relationship between kinetic energy and temperature of a gas gives us

$$\frac{3}{2}kT = \frac{1}{2}mv^2,$$

and solving for velocity gives

$$v = \sqrt{\frac{3\,kT}{m}}.$$

Here m refers to the mass of the particle whizzing around, in this case the proton. We insert this expression into the expression for the de Broglie wavelength to get

$$\lambda = \frac{h}{mv} = \frac{h}{\sqrt{3\,mkT}}.$$

Now we're ready to redo the calculation in part **a**, substituting the above expression for λ for r in our expression for temperature:

$$kT = \frac{qe^2}{3r} = \frac{qe^2\sqrt{3\,mkT}}{3h} = \frac{1}{\sqrt{3}}\frac{qe^2\sqrt{mkT}}{h}.$$

We need to solve for T. We divide each side by kT, and take the square, to find

$$T = \frac{1}{3}\frac{q^2e^4m}{h^2k}.$$

OK, let's plug in numbers:

$$T = \frac{1}{3}\frac{\left(9\times10^9\right)^2 \times \left(\frac{1}{6}\times10^{-18}\right)^4 \times \frac{1}{6}\times10^{-26}}{(\frac{2}{3}\times10^{-33})^2 \times \frac{1}{6}\times10^{-22}} \frac{\left(\frac{\text{Joule}\,\text{m}}{\text{Coulomb}^2}\right)^2 \times \text{Coulomb}^4 \times \text{kg}}{(\text{Joule}\,\text{sec})^2 \times \text{Joule}/\text{K}}.$$

We approximated Boltzmann's constant as $\frac{1}{6}\times10^{-22}$ in anticipation of some of the arithmetic to follow.

We've put all the units at the end. Let's work on the units first: note that all the Coulombs cancel; good. Next notice that the numerator has two factors of Joules, and the denominator has three of them, leaving us with units of

$$\frac{\text{meter}^2\,\text{kg}}{\text{Joule}\,\text{sec}^2} \times \text{K}.$$

Remembering that a Joule is $\text{kg}\,\text{meter}^2/\text{sec}^2$, the whole first expression cancels, leaving K, namely, units of temperature. That makes sense!

Now let's do the arithmetic. We do this by collecting together all the numerical factors, and all the powers of 10:

$$\frac{1}{3} \times \frac{9^2 \times \left(\frac{1}{6}\right)^5}{\left(\frac{2}{3}\right)^2 \times \frac{1}{6}} \times 10^{18-72-26+66+22}\,\text{K},$$

which becomes

$$\frac{3}{64} \times 10^8 \, \text{K} \approx \frac{1}{20} \times 10^8 \, \text{K} = 5 \times 10^6 \, \text{K},$$

which is a factor of 3 less than the true value of the temperature in the core of the Sun. In fact, it is very close to the core temperature of the coolest M dwarfs (i.e., those just barely massive enough for thermonuclear fusion to take place in their cores). We conclude that the temperature in the interior of the Sun is high enough for the protons to fuse.

This calculation was first done by the great Russian-American physicist George Gamow in the early 1930s, when astronomers were struggling to understand the processes by which stars shine. The process whereby two particles can fuse when coming within a de Broglie wavelength of each other is called *quantum tunneling*, and it comes up in discussions of the nature of the early universe in chapter 23 of *Welcome to the Universe*.

46. Properties of white dwarfs

46.a The parallax of a star in arcseconds is related to its distance in parsecs: distance = 1/parallax. Given the quoted parallax, we get a distance of 2.6 parsecs, or 8.6 light-years.

46.b The brightness of the Sun and Sirius B differ from each other because of their different distances and different luminosities. It is the ratio of luminosities we want. We now know the distance to Sirius B, so let's set up the relationships between brightness and luminosity:

$$b_{\text{Sirius B}} = \frac{L_{\text{Sirius B}}}{4\pi d_{\text{Sirius B}}^2},$$

$$b_{\text{Sun}} = \frac{L_{\text{Sun}}}{4\pi d_{\text{Sun}}^2}.$$

We take the ratio of these two equations and solve for the ratio of luminosities:

$$\frac{L_{\text{Sirius B}}}{L_{\text{Sun}}} = \frac{b_{\text{Sirius B}}}{b_{\text{Sun}}} \left(\frac{d_{\text{Sirius B}}}{d_{\text{Sun}}} \right)^2.$$

We know all the relevant quantities here, so let's see what we get:

$$\frac{L_{\text{Sirius B}}}{L_{\text{Sun}}} = 10^{-13} \times \left(\frac{2.6 \, \text{pc} \times 2 \times 10^5 \, \text{AU/pc}}{1 \, \text{AU}} \right)^2 = 10^{-13} \times 25 \times 10^{10} = 0.025.$$

Here we converted the distance to Sirius B to AU. Sirius B has a luminosity 1/40 that of the Sun.

46.c The white dwarf follows the Stefan-Boltzmann law, and therefore its temperature is related to its surface area and luminosity in the same sense as for the Sun:

$$L_{\text{Sirius B}} = \sigma T_{\text{Sirius B}}^4 \times 4\pi R_{\text{Sirius B}}^2,$$

$$L_{\text{Sun}} = \sigma T_{\text{Sun}}^4 \times 4\pi R_{\text{Sun}}^2.$$

Here σ is the Stefan-Boltzmann constant, whose value we will not need to know, as we'll see in a moment.

We wish to find the radius of Sirius B. We can progress by taking the ratio of these two equations:

$$\frac{L_{\text{Sirius B}}}{L_{\text{Sun}}} = \left(\frac{T_{\text{Sirius B}}}{T_{\text{Sun}}}\right)^4 \left(\frac{R_{\text{Sirius B}}}{R_{\text{Sun}}}\right)^2,$$

or

$$R_{\text{Sirius B}} = R_{\text{Sun}} \left(\frac{L_{\text{Sirius B}}}{L_{\text{Sun}}}\right)^{1/2} \left(\frac{T_{\text{Sirius B}}}{T_{\text{Sun}}}\right)^{-2}.$$

But we know all the ratios on the right-hand side of the equation: the ratio of temperatures is $25,000\text{K}/5,800\text{K} = 4.3$, and the ratio of luminosities is 0.025, so the radius of Sirius B is

$$R_{\text{Sirius B}} = R_{\text{Sun}} \times \frac{0.025^{1/2}}{18} = 0.009 \times R_{\text{Sun}} = 9 \times 10^{-3} \times 7 \times 10^5 \,\text{km} \approx 6,000 \,\text{km}.$$

Sirius B is just about the same size as Earth.

46.d We know the mass and the radius, so the density $\rho = M/V$ is straightforward:

$$\rho_{\text{Sirius B}} = M_{\text{Sirius B}}/V_{\text{Sirius B}} = \frac{M_{\text{Sun}}}{4\pi R_{\text{Sirius B}}^3/3},$$

$$\rho_{\text{Sirius B}} = \frac{2 \times 10^{30} \,\text{kg}}{4 \times (6 \times 10^6 \text{m})^3} \approx \frac{2 \times 10^{30} \,\text{kg}}{10^{21} \,\text{m}^3} = 2 \times 10^9 \,\text{kg m}^{-3},$$

or 2 million times the density of water (1,000 kilograms per cubic meter)! Astronomy involves extremes, from the extremely low density of interstellar and intergalactic space to the incredibly high density you've just calculated. In a future problem, you will examine the properties of neutron stars, which are higher density yet.

Let us consider a cubical nugget of side of length L and mass 1 ton $= 10^3$ kg; its volume is given by $V = M/\rho$, so

$$L = \left(\frac{10^3 \,\text{kg}}{2 \times 10^9 \,\text{kg m}^{-3}}\right)^{1/3} = (5 \times 10^{-7} \text{m}^3)^{1/3} = 8 \times 10^{-3} \,\text{m}.$$

Thus the nugget is just under a centimeter on a side; it will easily fit into a matchbox.

47. Squeezing into a white dwarf

Consider two carbon nuclei separated by a distance d. We may consider each carbon nucleus as occupying a volume of a cube of side d. The corresponding density is then the mass of the carbon nucleus divided by the volume of that cube. But the white dwarf has uniform density; the density of any small piece is equal to that of the white dwarf overall. This is enough information to calculate d:

$$\text{Density} = 2 \times 10^9 \,\frac{\text{kg}}{\text{m}^3} = \frac{12 \times \frac{1}{6} \times 10^{-26} \,\text{kg}}{d^3}.$$

Let's solve for d. A little rearranging gives

$$d^3 = 10^{-35} \,\text{m}^3 = 10 \times 10^{-36} \,\text{m}^3.$$

This form makes it easier to take the cube root. We know that $2^3 = 8$, which is pretty close to 10, so the cube root of 10 is about 2; certainly good enough for one significant figure! We then get

$$d = 2 \times 10^{-12} \, \text{meter}.$$

A rough number for the size of atoms under normal conditions (i.e., here on Earth) is about 1 Ångstrom; that is, about 10^{-10} meters. The carbon nuclei in a white dwarf are about 50 times closer together than that! We know that most of the volume of an atom under normal conditions is empty space; in a white dwarf, much of this empty space is squeezed out, and the nuclei are much closer together.

48. Flashing in the night

48.a The speed is the distance divided by the time. A point on the equator of the white dwarf goes all the way around ($2\pi r$) once in a period P, so the speed of that point is $v = 2\pi r/P$. Plugging in numbers gives

$$v = \frac{2 \times \pi \times 6.4 \times 10^6 \, \text{m}}{1/600 \, \text{sec}} \approx 2 \times 10^{10} \, \text{m/sec}.$$

Now, the speed of light is "only" 3×10^8 m/sec, so it is physically impossible for an object as large as a white dwarf to spin this fast; it would require that its surface travel faster than the speed of light!

48.b Going through the same calculation for the neutron star, we obtain

$$v = \frac{2\pi r}{P} = \frac{2 \times \pi \times 10^4 \, \text{m}}{1/600 \, \text{sec}} \approx 4 \times 10^7 \, \text{m/sec}.$$

This is incredibly fast but is still less than the speed of light, so it is physically possible.

48.c The centripetal acceleration is

$$a_c = v^2/r = \frac{(4 \times 10^7 \, \text{m/sec})^2}{10^4 \, \text{m}} = \frac{16 \times 10^{14}}{10^4} \, \text{m/sec}^2 = 1.6 \times 10^{11} \, \text{m/sec}^2,$$

or 16 billion times the acceleration of gravity on Earth's surface!

Can gravity supply the necessary acceleration? The acceleration of gravity on the neutron star surface is

$$g = \frac{GM_{\text{neutron star}}}{r^2} = \frac{\frac{2}{3} \times 10^{-10} \, \frac{\text{m}^3}{\text{sec}^2 \, \text{kg}} \times 2 \times 2 \times 10^{30} \, \text{kg}}{(10^4 \, \text{m})^2} = 3 \times 10^{12} \, \text{m/sec}^2.$$

The acceleration of gravity is greater than the centripetal acceleration, so it is adequate to keep you on the neutron star. Note that if the neutron star were spinning another factor of 3 or 4 faster, this would no longer be the case; even its own tremendous gravity would not be enough to hold it together, and the star would fly apart. Thus astronomers are not surprised that no pulsars have been found yet (despite intensive searching) that are spinning much faster than 600 times a second. The current record is a neutron star that spins 716 times a second; it lies in a globular cluster called Terzan 5.

49. Life on a neutron star

49.a This gravity is brutal. We first need to calculate the acceleration of gravity on the neutron star:

$$g = \frac{GM_{\text{neutron star}}}{r^2} = \frac{\frac{2}{3} \times 10^{-10} \frac{\text{m}^3}{\text{sec}^2\,\text{kg}} \times 2 \times 2 \times 10^{30}\,\text{kg}}{(10^4\,\text{m})^2} = 3 \times 10^{12}\,\text{m/sec}^2.$$

To lift up a mass $m = 1$ gram 1 centimeter requires an amount of energy

$$mgh = 1\,\text{g} \times 3 \times 10^{14}\,\text{cm/sec}^2 \times 1\,\text{cm} = 3 \times 10^{14}\,\text{erg} = 3 \times 10^7\,\text{Joules}.$$

That's a lot of energy! Our typical diet gives us $2000\,\text{cal} \times 4 \times 10^3\,\text{Joules/cal} = 8 \times 10^6\,\text{Joules}$ per day, so this represents 4 days of eating to gather the energy to lift this mass! Lifting up our own bodies (e.g., standing) is of course out of the question on a neutron star.

49.b We need to calculate the difference in gravitational acceleration g between two points, one just above the surface of the neutron star (i.e., a distance from the center of r_1, the radius of the neutron star), and the other a little bit farther away (a distance r_2; the difference between r_1 and r_2 is your height, say, 2 meters, much smaller than either r_1 or r_2).

So

$$\Delta g = \frac{GM}{r_1^2} - \frac{GM}{r_2^2} = GM\left(\frac{1}{r_1^2} - \frac{1}{r_2^2}\right).$$

Here M is the mass of the neutron star.

Let's keep on going, following the general rule that it is always a good thing to do algebra as long as possible before plugging in numbers:

$$\frac{1}{r_1^2} - \frac{1}{r_2^2} = \frac{r_2^2 - r_1^2}{r_1^2 r_2^2} = \frac{(r_2 - r_1)(r_2 + r_1)}{r_1^2 r_2^2}.$$

At this point, we can think about doing some approximations. We know that $r_2 - r_1 = 2$ meters, so that's easy. But we also know that the difference between r_1 and r_2 is quite small compared to either one of them, so to a very good approximation, $r_1 + r_2 \approx 2r_1$; similarly, $r_2^2 \approx r_1^2$. Plugging in these approximations, we find

$$\frac{1}{r_1^2} - \frac{1}{r_2^2} \approx \frac{(r_2 - r_1) \times 2r_1}{r_1^4} = \frac{2(r_2 - r_1)}{r_1^3}.$$

So we can now write

$$\Delta g = \frac{2GM(r_2 - r_1)}{r_1^3} = \frac{2GM\,\Delta r}{r_1^3},$$

where $\Delta r = r_2 - r_1$. What we've done, in essence, is to take a derivative here, without calling it that. You have just done some differential calculus! We can be cleverer still, and remember the expression for the acceleration itself:

$$g = \frac{GM}{r_1^2}.$$

So we can rewrite the difference as

$$\Delta g = 2\frac{\Delta r}{r}g,$$

which is useful, because we already calculated g in part **a**. Here we're asking for the difference in acceleration between our head and feet. For a height of 2 meters, this gives

$$\Delta g = 2 \times \frac{2\,\mathrm{m}}{10^4\,\mathrm{m}} \times 3 \times 10^{12}\,\mathrm{m/sec^2} \approx 10^9\,\mathrm{m/sec^2}.$$

So there is a stretching between your head and feet that's 100 million times stronger than the pull of gravity on Earth. You would be torn to little shreds before you hit the surface.

Another way to do this problem would be to simply calculate the acceleration of gravity at 10 km, and at 10 km + 2 meters, and take their difference. This will work, if you use a sufficient number of significant figures on your calculator. However, if we were asked to calculate the difference in acceleration over, say, 1 millimeter, most calculators could not handle it.

50. Distance to a supernova

50.a We are told that the Crab Nebula is approximately spherical and that it is expanding uniformly in all directions. In that case, the velocity of the outer shell will be uniform, 1,200 km sec^{-1}. This causes the diameter to increase at a rate of 0.23 arcseconds per year, corresponding to a rate of change of the distance from the center to the edge of half that, 0.12 arcseconds per year (as both "sides" of the shell are expanding away from the center). In 1 year, the outer shell moves a distance

$$1,200\,\mathrm{km\,sec^{-1}} \times 3 \times 10^7\,\mathrm{sec\,yr^{-1}} = 3.6 \times 10^{10}\,\mathrm{km}.$$

That corresponds to 0.12 arcseconds at the distance d of the supernova remnant, so we can use the small-angle approximation:

$$0.12\ \mathrm{arcsec} \times \frac{1\,\mathrm{radian}}{200,000\ \mathrm{arcsec}} = \frac{3.6 \times 10^{10}\ \mathrm{km}}{d}.$$

Solving for d gives

$$d = 3.6 \times 10^{10}\,\mathrm{km} \times \frac{2 \times 10^5}{0.12} = 6 \times 10^{16}\,\mathrm{km} \times \frac{1\ \mathrm{ly}}{10^{13}\,\mathrm{km}} = 6,000\,\mathrm{ly}.$$

50.b If we assume that the expansion rate has been constant throughout the history of the supernova, then it has been expanding for a time given by its current size divided by the rate of expansion (a calculation just like the time it takes to travel a certain distance at a certain speed). Thus the time the supernova has been expanding is

$$\frac{(5 \pm 1.5)\ \mathrm{arcmin} \times 60\,\mathrm{arcsec/arcmin}}{1/4\,\mathrm{arcsec/yr}} \approx 1200 \pm 300\,\mathrm{yr}.$$

We conclude that the light from the supernova reached us between around 500 and 1100 AD (i.e., between 900 and 1,500 years ago). The observations of the supernova in 1054 AD are consistent with the latter date.

Of course, the supernova is at a distance of 6,000 light-years, and therefore we are seeing events there as they occurred 6,000 years ago. It is the same 6,000-year delay between different events in the history of the supernova. This is why we said that the light from the supernova explosion reached us between 500 and 1100 AD. The supernova actually exploded 6000 years before that.

50.c The kinetic energy of a mass m moving at speed v is $\frac{1}{2}mv^2$, so the total amount of energy is:

$$E = \frac{1}{2}mv^2 = \frac{1}{2} \times 20\,M_{\text{Sun}} \times \frac{2 \times 10^{30}\,\text{kg}}{1\,M_{\text{Sun}}} \times (1.2 \times 10^6\,\text{m/s})^2$$

$$= 3 \times 10^{43}\,\text{Joules},$$

where we converted to MKS units. The luminosity of an O star, we are told, is 10^3 times that of the Sun, or roughly 4×10^{29} Joules/second. Thus the time the O star has to shine to generate that much energy is the ratio of the two numbers:

$$\text{Time} = \frac{3 \times 10^{43}\,\text{Joules}}{4 \times 10^{29}\,\text{Joules/sec}} = \frac{3}{4} \times 10^{14}\,\text{sec} \times \frac{1\,\text{yr}}{3 \times 10^7\,\text{sec}} = 2.5\,\text{million years}.$$

This is comparable to the main sequence lifetime of an O star.

51. Supernovae are energetic!

51.a We're told that the energy the supernova releases (in a month) is equal to that of the Sun in 10^8 years. We know the luminosity (energy released per unit time) of the Sun (namely 4×10^{26} Joules/sec), and so the total energy it releases is that luminosity times the time:

$$\text{Total energy} = 4 \times 10^{26}\,\frac{\text{Joules}}{\text{sec}} \times 10^8\,\text{years} \times 3 \times 10^7\,\frac{\text{sec}}{\text{year}} \approx 10^{42}\,\text{Joules}.$$

51.b We could use the above total energy and divide by the time of 1 month, but a simpler calculation is to compare with the Sun. We are told that the energy emitted by the supernova in a month is equal to that from the Sun in 10^8 years. And this energy is the luminosity times the time, so we find that

$$L_{\text{supernova}} \times 1\,\text{month} = L_{\text{Sun}} \times 10^8\,\text{years}.$$

Therefore, $L_{\text{supernova}} = 10^9 L_{\text{Sun}}$, where we've approximated 10 months per year (definitely good enough for one significant figure). Wow!

51.c We know the relationship between brightness b, distance d, and luminosity L:

$$b = \frac{L}{4\pi d^2}.$$

We are asked for the distance from the supernova at which it has the same brightness as the Moon. Thus let's equate the above expression for the Moon (for which $L = \frac{1}{4 \times 10^5} L_{\text{Sun}}$) and the supernova:

$$\frac{1}{4 \times 10^5} \frac{L_{\text{Sun}}}{4\pi d_{\text{Sun}}^2} = \frac{L_{\text{supernova}}}{4\pi d_{\text{supernova}}^2}.$$

Here d_{Sun} is our distance from the Sun (i.e., 1 AU). Before plugging in numbers, let's do a bit more algebra; we'll see that this will save us some trouble in a bit. Solving for our unknown (namely the distance to the supernova) gives us:

$$d_{\text{supernova}} = 1\,\text{AU} \times \left(\frac{4 \times 10^5 L_{\text{supernova}}}{L_{\text{Sun}}}\right)^{1/2}.$$

But we already know that the ratio of supernova to solar luminosity is 10^9. Thus

$$d_{\text{supernova}} = 1\,\text{AU} \times \left(4 \times 10^5 \times 10^9\right)^{1/2} = 2 \times 10^7\,\text{AU}.$$

There are 200,000 AU in a parsec, so this corresponds to a distance of 100 parsecs, or about 300 light-years.

52. Supernovae are dangerous!

The sphere of radius 30 parsecs has a volume

$$V = \frac{4\pi r^3}{3} \approx 4 \times (30\,\text{pc})^3 \approx 10^5\,\text{pc}^3.$$

Thus this volume contains roughly 100,000 stars, representing a fraction 10^{-6} of all the $\sim 10^{11}$ stars in the Milky Way. With one supernova going off in the entire Milky Way every 100 years, we will have to wait

$$\frac{100\,\text{years}}{10^{-6}} = 100\,\text{million years}$$

before one goes off close enough to cause damage. This timescale is short enough that it is entirely possible that the 3-plus billion-year history of the evolution of life on Earth has been affected multiple times by the explosion of a nearby supernova.

Note the large difference between the length of time calculated above and the very short lifetimes of stars that undergo a supernova (a few million years). The point here is that because of the various random motions of the stars in our Galaxy, the collection of stars that lie within 30 parsecs of the Sun will change with time. We know with confidence that there are no stars currently within 30 parsecs of the Sun that are massive enough to go supernova. However, as time goes on, as stars move around, it is very possible that a very massive star will come close to us and then go supernova.

53. Neutrinos coursing through us

53.a We have all the information we need to calculate the gravitational potential energy:

$$\text{Total energy emitted} = \frac{GM^2}{r} = \frac{\frac{2}{3} \times 10^{-10}\,\frac{\text{m}^3}{\text{sec}^2\,\text{kg}} \times (2 \times 10^{30}\,\text{kg})^2}{10^4\,\text{m}} \approx 3 \times 10^{46}\,\text{Joules}.$$

53.b If each neutrino has an energy of 1.5×10^{-12} Joules, the total number of neutrinos emitted by the star is straightforward:

$$3 \times 10^{46}\,\text{Joules} \times \frac{1\,\text{neutrino}}{\frac{3}{2} \times 10^{-12}\,\text{Joules}} = 2 \times 10^{58}\,\text{neutrinos}.$$

53.c These neutrinos are emitted essentially all at once, and thereafter, traveling at the speed of light, they expand into a huge spherical shell of ever-increasing radius. Thus by the time they impinge on Earth, they are spread out over a spherical shell of radius 150,000 light-years. The number density on the shell is

$$\frac{\text{Number}}{\text{Surface area}} = \frac{2 \times 10^{58}\,\text{neutrinos}}{4\pi(1.5 \times 10^5\,\text{ly} \times 10^{16}\,\text{m/ly})^2} \approx \frac{2 \times 10^{58}\,\text{neutrinos}}{2 \times 10^{43}\,\text{m}^2} = 10^{15}\,\text{neutrinos/m}^2.$$

That is, every square meter on Earth's surface was peppered with 1 quadrillion neutrinos from the supernova!

The neutrino detector is 10 meters on a side and therefore has a surface area of 100 square meters. Therefore 10^{17} neutrinos passed through it.

53.d We know how many neutrinos passed through Kamiokande, and we know that one in 5×10^{15} will be detected. Therefore, the total number of neutrinos detected is simply

$$10^{17} \text{ neutrinos} \times \frac{1 \text{ neutrino detected}}{5 \times 10^{15} \text{ neutrinos}} = 20 \text{ neutrinos}.$$

In fact, Kamiokande detected 11 neutrinos from the supernova. We got within a factor of 2 of this number with this very crude estimate. The Kamiokande detection remains the only time that neutrinos have been detected from any astronomical object more distant than the Sun.

It used to be thought that neutrinos, like photons, have no rest mass. That is, like photons, they travel only at the speed of light. It was discovered around the turn of the century (using the same Kamiokande experiment described above) that they in fact do have a very small mass, although the exact value is not yet known. Each neutrino has a mass less than 2×10^{-7} that of an electron. This discovery (together with the detection of neutrinos from the supernova) was recognized by the 2002 Physics Nobel Prize. John Bahcall, mentioned above, was one of the people responsible for this discovery, but was passed over for the prize...

54. A really big explosion

54.a Let's examine the information we're given. We know the rate at which photons hit BAT. We know how much energy each photon has, so we can convert from photons per second to Joules per second. The fact that this brightness was sustained for 30 seconds is irrelevant for this part of the problem. We also know the size of the BAT detector, so we can turn these numbers into a brightness (energy per unit time per unit area) by dividing by the area of the detector. So the brightness is given by

$$\text{Brightness} = \frac{\text{Photons per second on the detector} \times \text{Energy per photon}}{\text{Area of detector}}.$$

Plugging in numbers gives

$$\text{Brightness} = \frac{7 \times 10^4 \text{ photons/sec} \times 100 \text{ keV/photon} \times 1.6 \times 10^{-16} \text{ Joules/keV}}{0.05 \text{ m}^2},$$

where we have converted from square centimeters to square meters in the denominator.

We can do the arithmetic without a calculator, remembering that a single significant figure is just fine here. Indeed, to a single figure, $7 \times 1.6/5 = 2$, so we get

$$\text{Brightness} = 2 \times 10^{4+2-16+2} \frac{\text{Joules}}{\text{m}^2 \text{ sec}} = 2 \times 10^{-8} \frac{\text{Joules}}{\text{m}^2 \text{ sec}}.$$

54.b This calculation is straightforward; we simply take the ratio of the answer we got in part **a** to the number given here:

$$\frac{\text{Brightness in gamma rays}}{\text{Brightness in visible light}} = \frac{2 \times 10^{-8} \text{ Joules/sec/m}^2}{10^{-10} \text{ Joules/sec/m}^2} = 200.$$

This object was stunningly bright. We'll see in the next part that this object is about 11 billion light-years away. To have such a distant object appear bright enough to be seen by the naked eye is completely unprecedented in the history of astronomy. And yet the brightness in gamma rays was a factor of 200 higher still!

54.c We know the brightnesses b and distance d of the gamma-ray burst, and we need to calculate the luminosities (energy per unit time) and then multiply the result by the length of time t we see the explosion last (to get units of energy). This is a simple exercise in the inverse square law. Let's first do the calculation for the visible-light energy:

$$\text{Energy} = \text{Luminosity} \times \text{time} = 4\pi d^2 b t \times (1 + \text{redshift}).$$

The distance here is 1.1×10^{10} light-years. There are roughly 10^{16} meters in a light-year, so $d \approx 10^{26}$ meters. Multiplying by the duration of the burst as seen by us gives

$$\text{Energy} = 4\pi(10^{26}\,\text{meters})^2 \times 10^{-10}\,\text{Joules/sec/m}^2 \times 30\,\text{sec} \times 2,$$

which is roughly

$$\text{Energy} \approx 80 \times 10^{52-10+1}\,\text{Joules} = 8 \times 10^{44}\,\text{Joules}.$$

Thus to one significant figure, the energy of the explosion is 10^{45} Joules. That's the visible-light energy. The brightness of the gamma-ray emission is a factor of 200 times larger, and thus so is its energy. The emitted gamma-ray energy is therefore 2×10^{47} Joules.

This calculation made an assumption that may not be correct in detail: we of course see the emission coming directly toward us, and we have assumed that the emission is *isotropic* (i.e., the same in all directions). This assumption is built into the 4π term in the inverse square law relating brightness and luminosity, which we used above. Some models for gamma-ray bursts suggest that they are in fact beamed, with their emission emitted only along narrow cones. We then would see a burst only if its emission happened to be directed toward us. There are other astrophysical situations in which emission from objects is focused into a narrow beam, so this isn't a crazy idea. One popular model for gamma-ray bursts is that they are emitted when a particularly massive star collapses to a black hole at the end of its life; indeed, some gamma-ray bursts have been observed to go off coincident with supernovae. But theoretical astrophysicists still have only a rudimentary understanding of how this collapse results in a burst of gamma rays, and the extent to which this burst would be emitted in a narrow beam.

54.d Well, if $E = mc^2$, then $m = E/c^2$. Given the energy we just calculated in part **c** for the gamma-ray burst, the mass converted is

$$m = \frac{2 \times 10^{47}\,\text{Joules}}{(3 \times 10^8\,\text{m/sec})^2} \approx \frac{2 \times 10^{47}\,\text{kg}\,\text{m}^2/\text{sec}^2}{10^{17}\,\text{m}^2/\text{sec}^2} = 2 \times 10^{30}\,\text{kg}.$$

But the mass of the Sun is 2×10^{30} kg, so this is equivalent to the complete conversion of the entire mass of the Sun into pure energy! This is a truly outstandingly large explosion! Again, if the gamma-ray burst emission is in fact beamed, then we have overestimated the energy.

54.e The time for the Sun to give off a given amount of energy is simply the energy (Joules) divided by the Sun's luminosity (Joules/sec).

$$\text{Time} = \frac{8 \times 10^{44}\,\text{Joules}}{4 \times 10^{26}\,\text{Joules/sec}} \times \frac{1\,\text{year}}{3 \times 10^7\,\text{sec}} = \frac{2}{3} \times 10^{44-26-7}\,\text{years} \approx 7 \times 10^{10}\,\text{years}.$$

The Sun would have to shine for 70 billion years (about 7 times its actual lifetime) to emit this much energy. And here we used the visible-light energy; again, the gamma-ray energy is a factor of 200 larger still!

55. Kaboom!

55.a The energy E of a photon is related to its frequency ν by

$$E = h\nu,$$

where h is Planck's constant. The frequency in turn is related to its wavelength λ by

$$\nu = \frac{c}{\lambda},$$

where c is the speed of light. So calculating the energy is straightforward:

$$E = \frac{hc}{\lambda} = \frac{2/3 \times 10^{-33}\,\text{Joules sec} \times 3 \times 10^8\,\text{m/sec}}{7 \times 10^{-12}\,\text{m}} = \frac{2}{7} \times 10^{-33+8+12} \approx 3 \times 10^{-14}\,\text{Joules}.$$

We are given the flux of photons per second through each square meter. We have to convert this to brightness (energy per second through each square meter), which requires determining the amount of energy per photon. But we just calculated that! So it is a matter of simple multiplication:

$$\text{Brightness} = \frac{\text{Number of photons}}{\text{sec m}^2} \times \frac{\text{Energy}}{\text{photon}} = 10^{11}\,\frac{\text{photons}}{\text{sec meter}^2} \times 3 \times 10^{-14}\,\frac{\text{Joules}}{\text{photon}}$$

$$= 3 \times 10^{-3}\,\frac{\text{Joules}}{\text{sec meter}^2}.$$

55.b We have just calculated the brightness b. We now have to calculate the luminosity L. The two are related by the inverse square law:

$$L = 4\pi d^2 b,$$

where d is the distance of the object. We are given this distance. So, plugging in numbers and being careful to convert from light-years to meters, we get:

$$L = 4\pi \times (5 \times 10^4\,\text{ly} \times 10^{16}\,\text{m/ly})^2 \times 3 \times 10^{-3}\,\frac{\text{Joules}}{\text{sec m}^2}.$$

The quantity in parentheses is 5×10^{20} meters, which, when squared, gives 25×10^{40} meters^2. Multiplying 25 and 4 gives 100, and multiplying 3 and π gives a factor of 10. So we have

$$L = 100 \times 10 \times 10^{40} \times 10^{-3}\,\text{Joules/sec} = 10^{40}\,\text{Joules/sec}.$$

Now, it puts out that much energy for all of 0.1 second, so the total energy is just the product of these two numbers:

$$E = Lt = 10^{39} \text{ Joules.}$$

How long would it take the Sun to produce the same amount of energy? We know the luminosity of the Sun (Joules per second), that is, the rate at which it emits energy, and the total amount of energy, 10^{39} Joules. The ratio of the total energy to the rate is the time. Let's calculate that ratio, remembering to convert from seconds to years:

$$\text{Time} = \frac{\text{Energy}}{\text{Rate}} = \frac{10^{39} \text{ Joules}}{4 \times 10^{26} \text{ Joules/sec}} \times \frac{1 \text{ year}}{3 \times 10^7 \text{ sec}} = 10^{39-26-7-1} \text{ years} = 10^5 \text{ years,}$$

where we approximated $3 \times 4 = 10$ (anything to make the arithmetic easier!). So the amount of energy this gamma-ray burst put out in 0.1 seconds is equivalent to that which the Sun puts out in 100,000 years.

55.c Just as for the relationship between brightness and luminosity, the photon flux is related to the rate at which photons are emitted by the inverse square law:

$$\text{Flux} = \frac{\text{Number of photons per second}}{4 \pi d^2}.$$

We can write this expression down for two cases, the one observed, at a distance of $d_1 = 50,000$ light-years and a flux of $f_1 = 10^{11} \text{ photons s}^{-1} \text{ m}^{-2}$, and another case, in which the flux is now $f_2 = 10^4 \text{ photons s}^{-1} \text{ m}^{-2}$, and the distance d_2 is to be determined. In these two cases, the rate at which photons are emitted is the same (it is the same object, after all). That is,

$$f_1 = \frac{\text{Photon rate}}{4 \pi d_1^2},$$

$$f_2 = \frac{\text{Photon rate}}{4 \pi d_2^2}.$$

Let's divide the first equation by the second:

$$\left(\frac{d_2}{d_1} \right)^2 = \frac{f_1}{f_2},$$

(notice that the photon rate dropped out!) or, solving for d_2:

$$d_2 = d_1 \sqrt{\frac{f_1}{f_2}}.$$

Let's now plug in numbers:

$$d_2 = 50,000 \, \text{ly} \times \sqrt{\frac{10^{11}}{10^4}} \approx 50,000 \, \text{ly} \times 3 \times 10^3 = 1.5 \times 10^8 \, \text{ly.}$$

Thus the Swift satellite should be able to detect similar bursts 150 million light-years away (i.e., in moderately distant galaxies).

Note that we did as much algebra as possible before plugging in numbers. As we've seen before, this is a good idea; the calculation would have gotten terribly messy if we had plugged in numbers earlier. We could also have worked directly from the inverse square law, without taking ratios of equations; this would have been fine, but it would have required us to convert the value of the luminosity calculated in part **b** to photons, which would have been more work than needed.

56. Compact star

The density is the mass divided by the volume, and we can calculate this in reference to a single atom. The mass stays the same. The initial volume is proportional to the distance between atoms cubed. The final volume is proportional to the distance between nuclei cubed in the neutron star, when the nuclei touch one another. Thus the volume has decreased by a factor

$$\left(\frac{10^{-10}\,\text{m}}{10^{-15}\,\text{m}}\right)^3 = (10^5)^3 = 10^{15},$$

and thus the density has increased by the same factor. Ordinary solid objects here on Earth have densities of order 1 gram per cubic centimeter. Neutron stars, therefore, have a density of 10^{15}, or 1 quadrillion, grams per cubic centimeter!

57. Orbiting a neutron star

The speed is the distance divided by the time. In this case, let's take the time as one period and the distance as the circumference of the orbit. This is an orbit around an object of the same mass as the Sun (the fact that the density of the star is enormously higher makes no difference), which means that Kepler's third law in its simplest form holds, and the period in years is given by the radius of the orbit to the 3/2 power, or 27 years. For Earth, we know that

$$\text{Speed} = \frac{2\pi \times 1\,\text{AU}}{1\,\text{year}} = 30\,\text{km/sec}.$$

Here we have to do the calculation

$$\frac{2\pi \times 9\,\text{AU}}{27\,\text{years}},$$

which is clearly $9/27 = 1/3$ of speed of Earth around the Sun, or 10 km/sec.

Here's another way to do the problem: the speed v of a planet in a circular orbit is the circumference of the orbit ($2\pi a$, where a is the radius of the orbit) divided by its period P. Kepler's third law, in its simple form, tells us that for orbits around a solar-mass star, and in units of AU and years:

$$v = \frac{2\pi a}{P} \propto \frac{2\pi a}{a^{3/2}} = \frac{2\pi}{\sqrt{a}}.$$

That is, the speed is inversely proportional to the square root of the radius of the orbit. In this problem, the radius is 9 AU, and thus the speed is 1/3 that of Earth orbiting the Sun, or 10 km/sec.

Note that even though the neutron star is enormously more dense than the Sun, the orbit of a planet around the neutron star is governed by Kepler's laws in exactly the same way as the orbits in our own solar system.

58. The Hertzsprung-Russell diagram

The Hertzsprung-Russell diagram is shown in figure 7.1 of *Welcome to the Universe*. It plots luminosities of stars on the y-axis against their surface temperatures on the x-axis; confusingly, the temperature axis is backward, going from hot to cool from left to right. The main sequence, where most of the stars in the night sky lie, is a band of stars from upper left (hot and luminous) to lower right (cool and dim). There are also stars, termed "red giants," in the upper right of the diagram, while white dwarfs lie in the lower left of the diagram.

Stars radiate (approximately) as blackbodies. This means that the luminosity L of a star is related to its radius R and its surface temperature T by

$$L = 4\pi R^2 \sigma T^4,$$

where σ is the Stefan-Boltzmann constant. This explains the trend along the main sequence, whereby higher-temperature stars are significantly more luminous. A red giant is aptly named: it has a large luminosity despite its low temperature, because it has a radius much larger than that of a main sequence star. The largest supergiants have radii of 5 AU or even larger. Similarly, white dwarf stars are much smaller than main sequence stars (typical radii of a few thousand kilometers) and thus are much less luminous than are main sequence stars of the same surface temperature.

Stars on the main sequence are burning hydrogen to helium in their cores. The more massive the star is (on the main sequence), the more efficiently it burns, and therefore the more luminous it is. Such stars tend to have very high surface temperatures. Even though massive stars have more hydrogen fuel to burn, they burn it with such vigor that they have very short lifetimes compared to less massive stars. The most massive stars on the main sequence, 50 to 100 solar masses, have lifetimes of only a few million years. However, low-mass stars, say 1/10 the mass of the Sun, have lifetimes of up to 10 trillion years (much older than the current age of the universe; every star in the universe with such a low mass is still on the main sequence!). For comparison, the Sun has a lifetime of about 10 billion (10^{10} years), and is currently 4.6 billion years old; it is half-way through its full main sequence life.

When a star burns all the hydrogen in its core to helium, hydrogen burning starts in a shell around the core, and helium burns to carbon and oxygen in the core. Through a complex series of steps, the star becomes significantly more luminous and larger—a red giant. For a star less massive than about 8 solar masses, eventually the outer parts of the star will be thrown off, leaving behind the hot dense core, a white dwarf, made up mostly of carbon and oxygen. More massive stars will continue to fuse heavier elements even after the core is pure carbon or oxygen, only stopping when the core is pure iron. At this point, the star will explode as a supernova, leaving behind either a neutron star or black hole in the center.

59. A rival to Pluto?

59.a Because Eris is in opposition, the trajectory of Earth's orbit is perpendicular to the line between Earth and Eris. As Earth moves on its orbit, the direction in which we look to Eris changes. To do calculations, we need to know how fast Earth travels around the Sun. This can be done simply by remembering that Earth travels the circumference of its orbit in 1 year; thus the speed is given by the distance divided by the time:

$$v \approx \frac{2\pi \times 1.5 \times 10^8 \, \text{km}}{3 \times 10^7 \, \text{sec}} = 30 \, \text{km/sec}$$

to one significant figure (we made the standard approximation that $\pi \approx 3$).

At this speed, how far does Earth move in 5 hours (18,000 seconds)? Rounding to 20,000 seconds, the answer is $l = 600,000$ kilometers. We can approximate this distance as being along a straight line; in 5 hours, the curvature of the orbit is negligible. The parallax diagram is a long skinny triangle, with the 600,000 km of Earth's path at the base, the distance d to Eris its length, and a 7.5 arcsecond angle at its apex. We could use fancy notions from trigonometry to solve this triangle (remember, it is d that we're after here), but it is far easier to use the small-angle formula. If we measure angles θ in radians, and if the angles are small (as they often are in astronomy), then we can use the very simple relationship between the long side d and short side l of long skinny triangles:

$$\theta = \frac{l}{d}.$$

OK, we're almost done. In our case, we know θ (7.5 arcseconds), and we know l (about 6×10^5 km). We want to solve for $d = l/\theta$. Converting the angle to radians, we find

$$7.5 \, \text{arcsec} = 7.5 \, \text{arcsec} \times 5 \times 10^{-6} \, \text{radians/arcsec} \approx 4 \times 10^{-5} \, \text{radians}.$$

So the distance r to Eris is therefore

$$d = \frac{l}{\theta} = \frac{6 \times 10^5 \, \text{km}}{4 \times 10^{-5}} \approx 1.5 \times 10^{10} \, \text{km}.$$

There are 1.5×10^8 km in an AU, so the distance to Eris is 100 AU. (For this problem, we can take the distance from the Sun to Eris, and from Earth to Eris, as essentially the same; they differ by only 1 AU.)

The semi-major axis of Pluto's orbit is 40 AU; Eris is quite a bit more distant. Interestingly, Eris is on a highly elliptical orbit, $e = 0.44$ (much larger eccentricity than any of the planets in the solar system) and is currently very close to aphelion (its greatest distance from the Sun, 97.7 AU). Its perihelion (its closest approach to the Sun, 37.9 AU) is actually inside the orbit of Pluto, but it won't be there for another 280 years, given its 560-year orbital period.

59.b The brightness of the Sun as seen at Eris is given by the inverse square law:

$$b = \frac{L}{4\pi d^2}.$$

This is the amount of energy per unit time *per unit area* received at Eris. The cross-sectional area of Eris is πr^2, and thus the total energy per unit time falling on Eris is the product of the brightness and that area:

$$L \frac{r^2}{4 d^2}.$$

But only a fraction A of that light is reflected; the rest is absorbed. So our final answer is

$$AL \frac{r^2}{4 d^2}.$$

59.c What we have calculated in part **b** is the luminosity (in reflected light, at least) of Eris. We are observing it a distance d away (again, we're taking the approximation that the distance from the Sun to Eris, and from Eris to Earth, are essentially the same). So its brightness b just follows from the inverse square law:

$$b = AL\left(\frac{r}{2d}\right)^2 \times \frac{1}{4\pi d^2} = AL\frac{r^2}{16\pi d^4}.$$

The brightness of a distant Kuiper Belt Object falls off as the inverse fourth power of the distance. This is why it took so long for astronomers to discover these distant objects—they are really faint!

We have omitted are some details here. In particular, at any given time, only half the face of Eris is illuminated, and therefore it does not radiate its reflected light isotropically into a full sphere. Taking this into account properly requires a model for the detailed way in which light reflects from the surface, which is beyond the scope of this problem.

59.d To do this, let's solve our equation for r:

$$r = 4d^2\sqrt{\frac{\pi b}{AL}}.$$

Let's plug in numbers. We have to be careful to consistently use MKS units here; $d = 100\,\text{AU} = 1.5 \times 10^{13}$ meters:

$$r = 4 \times (1.5 \times 10^{13})^2 \times \sqrt{\frac{\pi \times 2.4 \times 10^{-16}}{0.96 \times 4 \times 10^{26}}} \text{ meters.}$$

How we can do this without a calculator? We can approximate $0.96 = 1$, so the quantity under the square root is roughly $0.6 \times \pi \times 10^{-42} = 2.0 \times 10^{-42}$, and its square root is about 1.4×10^{-21}. So we have

$$r = 4 \times 2.25 \times 10^{26} \times 1.4 \times 10^{-21} \text{ meters.}$$

$4 \times 2.25 = 9$, and 9×1.4 is roughly 13. So we get a radius of 13×10^5 meters, or 1,300 kilometers. This is seriously large, just about the same as Pluto itself! In fact, the modern precise value for the radius of Eris is 1,163 kilometers, and for Pluto it is 1,187 kilometers, just a bit bigger.

This calculation depended on us knowing the value of Eris's albedo. This is calculated by measuring both its brightness in visible light (reflected light from the Sun, proportional to A) and its infrared light (blackbody light due to the sunlight it has absorbed, proportional to $1 - A$). The comparison of the two allows us to determine both the size (as we have done here) and the albedo.

The discovery of Eris (named after the goddess of strife and discord in Greek mythology) set off a controversy in the astronomical community about whether it should be called a planet, and what the definition of a planet is. Reams have been written on this subject. The basic problem is that the concept of a planet has evolved as we have learned more, and we now realize that things that might conceivably be called planets now fall into a variety of categories:

- Terrestrial planets: relatively small rocky objects in the inner solar system, including Mercury, Venus, Earth, and Mars.

- Gas giants: much much larger and more massive bodies in the outer part of the solar system, including Jupiter, Saturn, Uranus, and Neptune. Note that Uranus and Neptune, which are considerably less massive than Jupiter and Saturn, actually mostly consist of frozen water, ammonia, and methane.
- Asteroids: objects mostly found in the main asteroid belt between Mars and Jupiter. They come in a large range of sizes. The largest of the objects in this belt is Ceres, often classified as a dwarf planet, with a diameter of 945 kilometers.
- Kuiper Belt Objects: icy bodies orbiting in the outer solar system. The largest of these include Pluto and Eris, and are also classified as dwarf planets.

And this list doesn't yet include the massive moons of Earth, Jupiter, and Saturn (the largest of which are considerably larger than Pluto) or the planets discovered around other stars. The term "planet" is now too broad to allow a single, all-encompassing clean definition, and our field has become richer with the discovery of Eris and its brethren.

59.e This is a simple application of Newton's form of Kepler's third law, which relates the period and semi-major axis of an orbiting body to the mass of the object it orbits:

$$P^2 = \frac{4\pi^2 a^3}{GM}.$$

We know the period. We could solve for the mass, if we knew the semi-major axis. What we're given is the angle that the semi-major axis subtends in our Hubble Space Telescope images. This is another opportunity to use the small-angle formula. Consider a very long skinny triangle, with length given by the distance from Earth to Eris (100 AU) and interior angle 0.53 arcsecond; we want to find the base of the triangle:

$$s = \theta d = 0.53\,\text{arcsec} \times \frac{1\,\text{radian}}{2 \times 10^5\,\text{arcsec}} \times 1.5 \times 10^{10}\,\text{kilometers} = 37{,}000\,\text{kilometers}.$$

Solving Kepler's third law for the mass yields

$$M = \frac{4\pi^2 a^3}{GP^2}.$$

We plug in the numbers to find

$$M = \frac{4\pi^2 \times (3.7 \times 10^7\,\text{m})^3}{2/3 \times 10^{-10}\,\text{m}^3\,\text{sec}^{-2}\,\text{kg}^{-1} \times (15.8\,\text{days} \times 86{,}400\,\text{sec/day})^2}.$$

Here we used a calculator to find a value of 1.6×10^{22} kg. This is remarkable: Eris is a bit more massive than is Pluto!

59.f The density is easy to calculate: it is simply the ratio of mass to volume. In fact, as we've found, Eris has a mass and volume both similar to that of Pluto, and so we will not be surprised to get a density similar to that of Pluto. But let's run the numbers, plugging in the mass and radius we have already calculated:

$$\rho = \frac{M}{V} = \frac{1.6 \times 10^{22}\,\text{kg}}{\frac{4}{3}\pi\,(1.3 \times 10^6\,\text{m})^3} = 1{,}800\,\text{kg/m}^3.$$

Now, there are 1,000 grams in a kilogram and 100 centimeters in a meter, and therefore a density of 1,000 kilograms per cubic meter can be converted as follows:

$$1\,\frac{\text{kg}}{\text{m}^3} = \frac{1{,}000\,\text{g}}{(10^2\,\text{cm})^3} = 10^{-3}\,\frac{\text{g}}{\text{cm}^3}.$$

Thus the density of Eris as we've calculated it is about 1.8 g/cm^3. To make a really careful comparison with Pluto, we would have to carry out our calculations to more significant figures (note throughout that we had to take cubes of a lot of numbers; this tends to magnify the uncertainties caused by rounding). Indeed, the modern value for the density of Eris is 2.5 grams per cubic centimeter. In any case, this density is somewhat higher than that of Pluto (1.9 g/cm^3), and higher than that of pure water ice. Therefore, we would guess that Eris is made of a mixture of rock and ice.

59.g Chapter 10 of *Welcome to the Universe* derives a formula for the equilibrium temperature of a planet around a star:

$$T_{\mathrm{p}} = T_* \left(\frac{r_*}{2d}\right)^{1/2} (1 - A)^{1/4},$$

where T_* and r_* are the surface temperature and radius of the star, d is its distance, and A is the albedo of the planet. Eris is remarkably reflective, with an albedo of 0.96; it absorbs only 4% of the light that impinges on it. So we expect it to be quite cold indeed. Plugging in numbers for the Sun and Eris, we find

$$T_{\mathrm{p}} = 6{,}000\,\text{K} \times \left(\frac{7 \times 10^5\,\text{km}}{2 \times 100\,\text{AU} \times 1.5 \times 10^8\,\text{km/AU}}\right)^{1/2} \times 0.04^{1/4}.$$

This gives a very chilly 10 K, to a single significant figure.

At what wavelength does the corresponding blackbody peak? The answer is given by Wein's law:

$$\lambda_{\mathrm{peak}} = \frac{0.3\,\text{cm}}{T} = \frac{3000\,\text{microns}}{10\,\text{K}} = 300\,\text{microns},$$

which is in fact in the far-infrared part of the spectrum, actually at somewhat longer wavelengths than Spitzer was able to observe. Nevertheless, observations with the Spitzer Space Telescope have been used to measure the blackbody spectrum from Eris and thus determine its temperature and its albedo.

59.h The question is whether the molecules are moving fast enough, given their temperature, to escape from the gravitational pull of Eris. Let us assemble what we will need. We need to do a calculation of the thermal speed of nitrogen molecules, given the temperature we've calculated. We also need to calculate the escape speed from the surface, which depends on Eris's radius r and mass M. The escape speed is given by $v_{\mathrm{esc}} = \sqrt{\frac{2GM}{r}}$, where G is Newton's gravitational constant. Let's plug in numbers:

$$v_{\mathrm{esc}} = \left(2 \times \frac{\frac{2}{3} \times 10^{-10}\,\text{m}^3\,\text{sec}^{-2}\,\text{kg}^{-1} \times 1.6 \times 10^{22}\,\text{kg}}{1.3 \times 10^6\,\text{m}}\right)^{1/2}.$$

Of course we want to do this without a calculator. Since $2 \times 2/3 \approx 1.3$, the 1.3 in the denominator cancels out. This gives $v_{\mathrm{esc}} = \sqrt{1.6 \times 10^6} \approx 1.3 \times 10^3$ meters per second.

Next, we need to calculate the typical speed of a nitrogen molecule. The kinetic energy of a nitrogen molecule, $\frac{1}{2}mv^2$, at a temperature T is $\frac{3}{2}kT$, where k is the Boltzmann constant. Thus, $v = \sqrt{\frac{3kT}{m}}$. A nitrogen molecule contains two atoms of nitrogen, each of which has an atomic mass of 14, so the mass of the molecule is 28 times that of a hydrogen atom. Using the atmospheric temperature we calculated above, we obtain a speed of

$$v = \sqrt{\frac{3 \times 1.4 \times 10^{-23}\,\text{Joules/K} \times 10\,\text{K}}{28 \times 1/6 \times 10^{-26}\,\text{kg}}}.$$

Again, let's try to evaluate this without a calculator. In the numerator $1.4 \times 10 = 14$, which cancels the 28 in the denominator, leaving a factor of 2 behind. The 1/6 in the denominator becomes a 6 in the numerator, which we multiply by 3 to give approximately 20, canceling the 2 in the denominator. Gathering the factors of 10, we find

$$v \approx 100\,\text{meters per second}.$$

So the escape speed from the surface is more than a factor of 10 larger than nitrogen's typical thermal speed; nitrogen will not escape. Indeed, at these low temperatures, it is likely to be in frozen form, and the icy surface is the reason the albedo is high.

The story of gas escaping from a planet is a bit more complicated than we've made it out to be here. The speed we've calculated for the nitrogen molecules is really an average, and the distribution of speeds is actually quite broad. That is, with a mean speed of 100 meters per second, some fraction of the nitrogen molecules are moving at speeds of 500 meters per second or larger. But in this case, the escape speed is a full factor of 10 larger than the average speed, so very few molecules are likely to escape.

60. Another Pluto rival

60.a As Earth moves, the direction to Orcus changes. To figure out what's going on, we need to know how far Earth moves in that time. We've seen before that the Earth travels around the Sun at 30 km/sec (we could calculate this again either from Kepler's third law, or by remembering that it travels around a complete circle of radius 1 AU in 1 year), so in 1 hour = 3,600 seconds, it travels a distance of

$$d = 30\,\text{km/sec} \times 3600\,\text{sec} = 1.08 \times 10^5\,\text{km}.$$

Note that the path Earth takes over this hour is very close to a straight line. The parallax diagram is a long skinny triangle, with the 108,000 kilometers of the Earth's path at the bottom, the distance r its length, and a 3.8 arcsecond angle at its apex. We will use the small-angle formula relating the angle θ in radians, the long side r and short side d of long skinny triangles:

$$\theta = \frac{d}{r}.$$

So, converting to radians gives

$$3.8'' = 3.8'' \times 5 \times 10^{-6}\,\text{radians}/'' = 1.9 \times 10^{-5}\,\text{radians}.$$

So the distance r to Orcus is therefore

$$r = \frac{d}{\theta} = \frac{1.08 \times 10^5 \, \text{km}}{1.9 \times 10^{-5}} \approx 6 \times 10^9 \, \text{km}.$$

There are 1.5×10^8 km in an astronomical unit, so the distance to Orcus is 40 AU. (For this problem, we can take the distance from the Sun to Orcus and from Earth to Orcus as essentially the same.)

The semi-major axis of Pluto's orbit is also 40 AU, so Orcus is in a similar orbit to Pluto's.

60.b This is a problem in Kepler's third law; it concerns an orbit around the Sun, so we can use the simple form of the law:

$$\left(\frac{P}{\text{years}} \right)^2 = \left(\frac{a}{\text{AU}} \right)^3 - 40^3 = 6.4 \times 10^4$$

$$P = \sqrt{6.4 \times 10^4} = 2.4 \times 10^2 \, \text{years} = 240 \, \text{years}.$$

The period is same as that of Pluto (as it must be, given that it has the same semi-major axis!). How about the speed? There are many ways we could do this; here's a clever way. We know that Earth goes around the Sun at 30 km/sec. We also know that the orbital speed a distance r from the Sun of mass M is $\sqrt{GM/r}$, and both are orbiting the Sun. So an object that is 40 times farther away from the Sun has a speed that is $\sqrt{40} \approx 6.5$ times slower (the M in the speed expression is that of the Sun for the orbits of both Earth and Orcus). That is, the speed of Orcus is $30/6.5 \approx 4.5$ km/sec. So our approximation that it was essentially standing still (relative to Earth) is a good one!

60.c Kepler's third law tells us that the period of Neptune's orbit is

$$P = \sqrt{30^3} = \sqrt{2.7 \times 10^4} = 160 \, \text{years}.$$

Let's take the ratio with Orcus's period: $240/160 = 1.5 = 3/2$. We say that Orcus is in a three-to-two resonance with Neptune. Every three orbits of Neptune, and every two orbits of Orcus, the two bodies find themselves in the same point of their orbits relative to each other, and they give each other a tug (Neptune, being the much more massive of the two, gives Orcus a larger acceleration than Orcus gives Neptune). These resonances are complicated; in some cases, these tugs would cause the planet to be moved out of its orbit. In other cases, such as this one, the resonance attracts bodies to it. Indeed, the TNO website quoted above contains roughly 1,800 objects as of February 2017. Of these, a full 390 (over 20%) have semi-major axes between 39 and 41 AU; they are all in this resonance.

So rather than thinking of Pluto as the wimpiest, least-massive planet, we should think it in far loftier terms as the King of the Kuiper Belt, the largest of the Plutinos.

60.d Let's start by figuring out how much energy per unit time Orcus receives from the Sun. It has a cross-sectional area of πR_p^2. This is the area of the shadow that Orcus casts, and so the fraction of all the energy the star puts out that Orcus receives is that cross-sectional area divided by the surface area of the sphere centered on the Sun out to Orcus:

$$\frac{\pi r^2}{4 \pi d^2} = \left(\frac{r}{2d} \right)^2.$$

Thus the amount of light per unit time incident on Orcus is $L(r/2d)^2$. Of the light incident on it, it reflects a fraction A, so the total amount of light reflected is: $AL(r/2d)^2$.

60.e What we have calculated in part **d** is the luminosity (in reflected light, at least) of Orcus. We are observing it from a distance d away (again, we're taking the approximation that the distances from the Sun to Orcus and from Orcus to Earth, are essentially the same). So its brightness just follows from the inverse square law:

$$\text{brightness } b = AL\left(\frac{r}{2d}\right)^2 \times \frac{1}{4\pi d^2} = AL\frac{r^2}{16\pi d^4}.$$

60.f To do this, let's solve our equation for r:

$$r = 4d^2\sqrt{\frac{\pi b}{AL}}.$$

Now plug in numbers. We have to be careful to consistently use MKS units here, so $d = 40\,\text{AU} = 6 \times 10^{12}$ meters:

$$r = 4 \times (6 \times 10^{12})^2 \times \sqrt{\frac{\pi \times 4 \times 10^{-16}}{0.23 \times 4 \times 10^{26}}}.$$

Let's see how we can do this without a calculator. The quantity under the square root is roughly 13×10^{-42}; its square root is about 3.6×10^{-21}. So we have

$$r = 4 \times 36 \times 10^{24} \times 3.6 \times 10^{-21}.$$

To do this without a calculator, remember that $36 = 9 \times 4$, so

$$4 \times 36 \times 3.6 = 0.1 \times 9^2 \times 4^3 \approx 8 \times 64 \approx 500.$$

So we have finally that

$$r = 500 \times 10^3\,\text{m} = 500\,\text{km}.$$

Thus the diameter of Orcus is 1,000 kilometers. This is seriously large; it is almost half the diameter of Pluto itself.

Orcus is named for the Etruscan god of the underworld. And it has been discovered that Orcus has a moon orbiting it, called "Vanth" (also a name from Etruscan mythology), which is roughly a third the diameter of Orcus.

61. Effects of a planet on its parent star

61.a The momenta of the planet and star are equal and opposite, and thus we have

$$M_* v_* = M_p v_p.$$

(Strictly speaking, there should be a minus sign in this equation, which simply is a statement that the planet and star move in opposite directions. We need not keep this minus sign in our calculations for the purposes of this problem.) Here we're asked to calculate the speed of the star

$$v_* = \frac{M_p}{M_*}\sqrt{\frac{GM_*}{r}} = M_p\sqrt{\frac{G}{rM_*}},$$

which is quite accurate if the planet's mass is much smaller than the star's mass (a valid assumption in practice!).

61.b Now we plug in numbers. Let's first calculate the speed of Jupiter around the Sun:

$$v_{\text{Jupiter}} = \sqrt{\frac{GM_{\text{Sun}}}{5\,\text{AU}}} = \sqrt{\frac{2/3 \times 10^{-10}\,\text{m}^3\,\text{sec}^{-2}\,\text{kg}^{-1} \times 2 \times 10^{30}\,\text{kg}}{5 \times \frac{3}{2} \times 10^{11}\,\text{m}}}$$

$$= \sqrt{\frac{16}{9} \times 10^8\,(\text{m/sec})^2}$$

$$= 4/3 \times 10^4\,\text{meters/sec} \approx 13\,\text{km/sec},$$

or a bit less than half the speed of Earth going around the Sun. So how large is the motion of the Sun? Well, the mass of Jupiter is very close to 1/1,000 that of the Sun, so from what we derived above, the Sun moves 1/1,000 as fast as Jupiter, or 13 m/sec. That's a bit faster than the 10 m/sec of the fastest runners, but not by much.

61.c Here we need to calculate the semi-major axis of the planet in each case. We're told that the star in question has a mass equal to that of the Sun, so we can use Kepler's third law in its simplest form, if we express the period in years:

$$a(\text{AU}) = P(\text{years})^{2/3}.$$

This is trivial for HD 156836; the period is very close to 1 year, so the semi-major axis is 1 AU. For 51 Pegasi, the period is very short; 4.2 days is equivalent to 0.011 years, and plugging this into Kepler's third law gives a semi-major axis of 0.05 AU. It is really close to its parent star.

Given the semi-major axis in the two cases, we can use the equation we derived above in part **a** relating the speed of the star to the mass of the planet. We rewrite it as

$$M_{\text{p}} = v_* \sqrt{\frac{rM_*}{G}}.$$

We've calculated the orbital radii r in each case. Indeed, we can scale directly off the mass of Jupiter (as we're asked to calculate the mass in Jupiter masses). That is, we can write, from our work in part **b**, that

$$M_{\text{Jupiter}} = v_{\text{Sun}} \sqrt{\frac{r_{\text{Jupiter}} M_{\text{Sun}}}{G}}.$$

Taking the ratio of these two equations and again recognizing that the mass of the star is that of the Sun, we find:

$$\frac{M_{\text{p}}}{M_{\text{Jupiter}}} = \frac{v_*}{13\,\text{km/sec}} \sqrt{\frac{r}{r_{\text{Jupiter}}}}.$$

When we plug in numbers for the two cases, the arithmetic is straightforward:

$$\text{For 51 Pegasi: } M_{\text{p}} = \frac{56\,\text{m/sec}}{13\,\text{m/sec}} \times \sqrt{\frac{0.05\,\text{AU}}{5\,\text{AU}}}\, M_{\text{Jupiter}} \approx 0.4\, M_{\text{Jupiter}}\,.$$

$$\text{For HD 156836: } M_p = \frac{464\,\text{m/sec}}{13\,\text{m/sec}} \times \sqrt{\frac{1\,\text{AU}}{5\,\text{AU}}}\, M_{\text{Jupiter}} \approx 15\, M_{\text{Jupiter}}\,.$$

Thus the 51 Pegasi planet has a mass a bit less than half that of Jupiter, while HD 156836 is 15 times more massive than Jupiter. Notice, incidentally, how different these are from our own solar system. The closest planet to our Sun, Mercury, is tiny compared to Jupiter, and is quite a bit farther out at 0.4 AU. And we certainly don't have any supermassive planets in our solar system, anywhere near 15 Jupiter masses!

Note that the values we've calculated are not quite right, simply because we made the simplifying assumption that the mass of the star in each case was exactly the same as that of the Sun. More importantly, the Doppler effect is only sensitive to motions along the line of sight to the star, and so the effect depends on the orientation of the orbit relative to the plane of the sky. Thus we observe only one component of the orbital motion, which means that the masses we have calculated are lower limits to the true values.

Finally, we need to calculate the surface temperatures of these planets. We're to assume no albedo and no greenhouse effect, so the equation is a simple one:

$$T_{\rm p} = T_* \sqrt{\frac{R_*}{2r}}.$$

Plugging in the surface temperature of the Sun and the orbital radii in the two cases, we find

$$\text{For 51 Pegasi: } T_{\rm p} = 6{,}000 \times \sqrt{\frac{7 \times 10^5 \,\text{km}}{2 \times 0.05 \times 1.5 \times 10^8 \,\text{km}}} \approx 1{,}200\,\text{K}.$$

That's seriously hot! For the planet that's farther out:

$$\text{For HD 156836: } T_{\rm p} = 6{,}000 \times \sqrt{\frac{7 \times 10^5 \,\text{km}}{2 \times 1.5 \times 10^8 \,\text{km}}} \approx 300\,\text{K},$$

pretty close to the average surface temperature of Earth. One might speculate indeed that this last planet might have liquid water, given this temperature. However, these planets have masses comparable to Jupiter or larger. Such massive planets are made mostly of hydrogen and helium, like the Sun. Thus they are gaseous and do not have a solid surface to harbor oceans.

We have already mentioned that the Doppler shift technique is only sensitive to motions along the line of sight. That is, the calculation of the masses is correct only if the plane of the orbit is perpendicular to the plane of the sky. If the orbit is tilted, then we have measured only a fraction of the speed of the star, and the masses we have calculated are lower limits. If the true mass is larger than 13 times the mass of Jupiter, the interior temperatures are high enough for deuterium nuclei (but not hydrogen nuclei) to undergo fusion. Such objects are classified as brown dwarf stars, not planets.

62. Catastrophic asteroid impacts

62.a The mass of the water M to a depth of $h = 200$ meters is given by the density of water ($\rho = 1{,}000$ gm/m^3) times the volume of this water. The volume is its area times its thickness:

$$\frac{3}{4} \times \text{Surface area of Earth} \times h,$$

giving a total mass of

$$M = 3\pi R^2 h\rho,$$

where R is the radius of Earth, and we used the expression for the surface area of a sphere, $4\pi R^2$. Plugging in numbers gives us a mass of

$$M = 3 \times \pi \times (6.4 \times 10^6 \, \text{m})^2 \times 200 \, \text{m} \times 1{,}000 \, \text{kg/m}^3.$$

Take $3 \times \pi = 10$ and $6.4^2 \approx 40$, so we get a total mass of $M = 8 \times 10^{19}$ kg in the upper 200 meters of Earth's oceans.

If each kilogram of that mass requires 2.5×10^6 Joules to boil (i.e., to vaporize and turn into steam), the total energy required to vaporize everything is then

$$8 \times 10^{19} \, \text{kg} \times 2.5 \times 10^6 \, \text{Joules/kg} = 2 \times 10^{26} \, \text{Joules}.$$

That's a seriously big number! (It is about the amount of energy the Sun gives off in half a second.)

62.b The expression for kinetic energy is $\frac{1}{2}mv^2$. We want the kinetic energy of the asteroid (in fact, 1/4 of that energy) to be enough to boil the water (i.e., to be equal to the energy we calculated in part **a**). We know the velocity v of the asteroid and are asked to determine its mass. Thus the expression becomes

$$\frac{1}{4} \times \frac{1}{2}mv^2 = 2 \times 10^{26} \, \text{Joules}.$$

(Here we are doing the calculation necessary for the top 200 meters of oceans; we'll do the full oceans below). Solving for m gives us:

$$m = \frac{8 \times 2 \times 10^{26} \, \text{kg} \, \text{m}^2/\text{sec}^2}{(2 \times 10^4 \, \text{m/sec})^2} = 4 \times 10^{18} \, \text{kg}.$$

Note that we have expressed the units of Joules in terms of kilograms, meters, and seconds. That is a big number indeed! We'll see just how big it is below.

For the full oceans, the numbers will be even bigger. The mean depth of the oceans is about 3.5 kilometers (i.e., almost 20 times the 200 meters of the calculation we did above). So the full mass of the oceans is a factor of 20 times larger than what we calculated above, and thus the kinetic energy, and so the mass, of the incoming asteroid needed to vaporize all the oceans will be $3{,}500/200$ times larger, namely, 7×10^{19} kg.

62.c The relationship between mass, density, and volume, as we've already seen, is $m = \rho V$. We have ρ and m, and are asked to calculate V. Solving for V is straightforward, but before we do so, let's remember that we've been given ρ in units that don't match; we need to convert to kilograms per cubic meter:

$$\rho = 3.0 \, \frac{\text{g}}{\text{cm}^3} \times \frac{1 \, \text{kg}}{1{,}000 \, \text{g}} \times \left(\frac{100 \, \text{cm}}{\text{m}} \right)^3 = 3{,}000 \, \frac{\text{kg}}{\text{m}^3}.$$

Now we're ready to calculate volumes. For the epipelagic-zone-vaporizing asteroid, we get

$$V = \frac{m}{\rho} = \frac{4 \times 10^{18} \, \text{kg}}{3{,}000 \, \text{kg/m}^3} \approx 1 \times 10^{15} \, \text{meters}^3,$$

to one significant figure. And for all the oceans, we similarly get

$$V = \frac{m}{\rho} = \frac{7 \times 10^{19}\,\text{kg}}{3000\,\text{kg/m}^3} \approx 2 \times 10^{16}\,\text{meters}^3.$$

62.d We have calculated the volumes of the asteroids in part **c**; now we need to figure out the radii r. For a sphere, the volume is $4\pi r^3/3$. So we want to solve for the radii. At the single significant figure we're doing calculations here, $\pi = 3$, so solving for radius gives

$$r = \left(\frac{V}{4}\right)^{1/3}.$$

Plugging this in for the two cases above gives

$$r = \left(\frac{10^{15}\,\text{meters}^3}{4}\right)^{1/3}$$

$$\approx 60,000\,\text{meters for the asteroid that vaporizes the epipelagic zone.}$$

This is indeed a huge asteroid, with a radius of 60 kilometers. A similar calculation for the asteroid that evaporates all the oceans gives a radius of 170 kilometers, even larger.

We are told that the number of asteroids impacting Earth of a given radius is proportional to the inverse square of that radius. Thus if there were three ocean-vaporizing asteroids in the period of Late Heavy Bombardment, then

$$3 \times \left(\frac{60\,\text{km}}{170\,\text{km}}\right)^{-2} = 20$$

asteroids (to one significant figure) were big enough to vaporize the epipelagic zone. That is, it is possible that if life got started in the early history of Earth, it was wiped out 20 times over!

If these impacts happened over a period of 600 million years, then the average time between impacts was 600 million years/20, or about 30 million years.

63. Tearing up planets

63.a Before starting, notice that we are slightly abusing Newton's second law of motion here. It states that the acceleration of an object times its mass is equal to the sum of all the forces acting on the object. Here, we want the acceleration due to each force individually; this is each individual force, divided by the relevant mass. The total acceleration is just the sum of the individual accelerations.

Consider two rocks, labeled 1 and 2, at the center and near surface of the moon. The distance each is from the center of the planet (which is from where the effective gravitational force acts) is r, and $r - R_m$, respectively. Newton's second law of motion and law of gravitation ensure that the acceleration on each rock due to the gravity of the planet is independent of the mass of the rock. These accelerations are given by

$$a_1 = \frac{GM_{\text{p}}}{r^2},$$

$$a_2 = \frac{GM_p}{(r - R_m)^2},$$

where M_p is of course the mass of the planet. We are not given this quantity, but we know the mass of the planet to be $M_p = \frac{4\pi}{3}\rho R_p^3$ (mass is equal to density times volume), so

$$a_1 = \frac{GM_p}{r^2} = \frac{4\pi G\rho R_p^3}{3r^2},$$

$$a_2 = \frac{GM_p}{(r - R_m)^2} = \frac{4\pi G\rho R_p^3}{3(r - R_m)^2}.$$

The problem asks for the difference in the two accelerations. The surface rock is closer to the planet and thus feels a larger gravitational tug. Thus we calculate

$$\Delta a = a_2 - a_1 = \frac{4\pi G\rho R_p^3}{3(r - R_m)^2} - \frac{4\pi G\rho R_p^3}{3r^2} = \frac{4\pi G\rho R_p^3}{3}\left[\frac{1}{(r - R_m)^2} - \frac{1}{r^2}\right].$$

Put these two fractions over a common denominator, and expand the square in the numerator:

$$\Delta a = \frac{4\pi G\rho R_p^3}{3}\frac{r^2 - (r^2 - 2rR_m + R_m^2)}{(r - R_m)^2 r^2}$$

$$= \frac{4\pi G\rho R_p^3}{3}\frac{2rR_m - R_m^2}{(r - R_m)^2 r^2}.$$

Let us rewrite this in a suggestive way, by factoring out all factors of r that we can in the numerator and denominator:

$$\Delta a = \frac{4\pi G\rho R_p^3}{3}\frac{rR_m\left(2 - \frac{R_m}{r}\right)}{r^4\left(1 - \frac{R_m}{r}\right)^2}.$$

The calculation is almost finished, but we haven't yet used the fact that $R_m \ll r$; this being the case, $\frac{R_m}{r} \ll 1$ and so we can ignore a few terms:

$$\Delta a = \frac{4\pi G\rho R_p^3}{3}\frac{2rR_m}{r^4} = \frac{8\pi G\rho R_p^3 R_m}{3r^3}.$$

The steps just carried out in fact use the methods of differential calculus, even though we have not called it that.

This difference in the acceleration happens because the rock that is closer to the planet will be pulled more strongly toward the planet than the rock farther away. Thus the effect of this difference is to work to pull these two rocks apart.

63.b These two rocks are held together by the self-gravity of the moon. The rock at the center feels no such force; the moon is arranged symmetrically around it, and thus does not pull more in one direction than in another. But the rock at the surface feels the full gravitational effect of the moon. This is an inward force, toward the central rock, and thus counteracts the stretching tendency of the tidal force we calculated in part **a**. The corresponding acceleration is also given by Newton's law of gravitation:

$$a_{\text{self-gravity}} = \frac{GM_m}{R_m^2} = \frac{4\pi G\rho R_m^3}{3R_m^2} = \frac{4\pi G\rho R_m}{3},$$

where M_m is the total mass of the moon.

63.c The two effects, the tidal force induced by the planet and the self-gravity of the moon, work in opposite senses. The moon will lose the struggle and will be torn apart by the tidal effects when the tidal force is greater than the self-gravity:

$$\frac{8\pi G \rho R_\mathrm{p}^3 R_\mathrm{m}}{3 r^3} > \frac{4\pi G \rho R_\mathrm{m}}{3}$$

$$\frac{2 R_\mathrm{p}^3}{r^3} > 1,$$

or

$$2 R_\mathrm{p}^3 > r^3$$

$$2^{1/3} R_\mathrm{p} = 1.26 R_\mathrm{p} > r.$$

Moons that orbit closer than this are likely to be torn apart by tidal forces.

63.d There are no moons of the rocky planets within the Roche limit of the planets, but there are a few of them for the gas giants (Metis and Adrastea around Jupiter; Pan, Atlas, Prometheus, and Pandora for Saturn; Cordelia, Ophelia, Bianca, and Cressida for Uranus; Naiad, Thalassa, and Despina for Neptune). Our calculation found the Roche limit to be independent of the density, but we did assume that the density of the planet and the moon are the same. If the density of the moon is higher than the (low) density of the gas giant, then the self-gravity will be stronger, allowing some rocky moons to survive inside the Roche limit of gaseous planets.

The rings of Saturn lie within the Roche limit of the planet and are made of particles of ice of a density similar to that of Saturn. These particles have not formed into moons because of the tidal forces due to Saturn. Icy moons *are* found, however, beyond the Roche limit in orbit around Saturn.

64. Planetary orbits and temperatures

64.a This problem is an exercise in Kepler's third law. This problem differs from that of Earth around the Sun because the star has a larger mass and Prefect has a larger semi-major axis. Let's write down Newton's form of Kepler's third law for the two cases:

$$P_\mathrm{Earth}^2 = \frac{4\pi^2 a_\mathrm{Earth}^3}{G M_\mathrm{Sun}},$$

$$P_\mathrm{Prefect}^2 = \frac{4\pi^2 a_\mathrm{Prefect}^3}{G M_\mathrm{star}}.$$

We take the ratio of the two. Lots of terms drop out:

$$\left(\frac{P_\mathrm{Prefect}}{P_\mathrm{Earth}}\right)^2 = \left(\frac{a_\mathrm{Prefect}}{a_\mathrm{Earth}}\right)^3 \left(\frac{M_\mathrm{Sun}}{M_\mathrm{star}}\right) = 4^3/4 = 4^2.$$

Therefore $P_\mathrm{Prefect} = 4 P_\mathrm{Earth}$, or 4 years.

64.b If we can ignore albedo and the greenhouse effect, the equation for the equilibrium temperature of the planet is

$$T_\mathrm{planet} = T_\mathrm{star} \sqrt{\frac{R_\mathrm{star}}{2 d}},$$

where R_{star} is the radius of the parent star, and d is the distance from the star to the planet. In this problem, we're comparing two planets around the same star, so T_{star} and R_{star} are the same. When we take the ratio, then, things are very simple:

$$\frac{T_{\text{Zaphod}}}{T_{\text{Prefect}}} = \sqrt{\frac{d_{\text{Prefect}}}{d_{\text{Zaphod}}}} = \sqrt{4} = 2.$$

64.c The density is the ratio of the mass to the volume:

$$\rho_{\text{Zaphod}} = \frac{M_{\text{Zaphod}}}{\frac{4}{3}\pi R_{\text{Zaphod}}^3},$$

$$\rho_{\text{Prefect}} = \frac{M_{\text{Prefect}}}{\frac{4}{3}\pi R_{\text{Prefect}}^3},$$

where we've recognized that the two are both spheres and used the appropriate expression for the volume of a sphere. We can take their ratio to find

$$\frac{\rho_{\text{Zaphod}}}{\rho_{\text{Prefect}}} = \left(\frac{R_{\text{Prefect}}}{R_{\text{Zaphod}}}\right)^3 = 2^3 = 8,$$

where we've recognized that the masses are identical and thus drop out.

65. Water on other planets?

65.a As the parent stars in each case are similar to the Sun, they have the same mass. Thus Kepler's third law is straightforward to apply in every case; if we convert the period to years, we can use Kepler's simple form of his third law (the semi-major axis in AU is equal to the period in years, to the 2/3 power) to determine the semi-major axis in AU.

Once we have the semi-major axis a (and we are given the eccentricity e), the distances of closest and farthest approach are also simple; they are $a(1-e)$ and $a(1+e)$, respectively. The table below gives all the relevant numbers (and yes, we used a calculator) in AU; in each case, we converted the period from days to years.

Name of planet	period (years)	a (AU)	Closest approach (AU)	Farthest approach (AU)
51 Pegasi	0.012	0.051	0.051	0.052
HD 209458	0.0097	0.045	0.040	0.050
55 Cancri b	0.040	0.12	0.11	0.12
55 Cancri c	0.12	0.24	0.16	0.33
55 Cancri d	15	6.0	5.0	7.0
HD 142415	1.1	1.04	0.52	1.56

Note that in each case, the results are given to two significant figures; for those planets with very small eccentricities, the difference between closest and farthest approach is very small.

65.b With zero albedo, the relationship between the temperature of the planet and its parent star is:

$$T_{\text{planet}} = T_{\text{star}}\left(\frac{R_{\text{star}}}{2d}\right)^{1/2}.$$

We can do this for each of the above planets, using the distances we've calculated and the properties of the Sun (again, each of these stars is assumed to be like the Sun). Using a calculator, we again make a table (all temperatures are in Kelvin):

Name of planet	Period (years)	Average T	T at closest approach	T at farthest approach
51 Pegasi	0.012	1,300	1,300	1,300
HD 209458	0.0097	1,400	1,400	1,300
55 Cancri b	0.040	850	850	840
55 Cancri c	0.12	580	720	500
55 Cancri d	15	120	130	110
HD 142415	1.1	280	400	230

Note that temperatures are given to two significant figures; more than this is inappropriate, both because the input numbers for the properties of the Sun are no better than that, and because the simplifying assumptions about the albedo and the greenhouse introduce large uncertainties.

65.c When the discovery of these planets was first made, astronomers around the world carried out exactly the calculation you have just done, to answer this question. One of the great surprises to come out of these observations was that, unlike our own solar system, there are very massive planets (these planets have typical masses of several times that of Jupiter) very close to their parent stars. And as you have just seen, these planets are *hot*. Even an extremely high albedo (unlikely) is not going to get the so-called hot Jupiters 51 Pegasi, HD 209458, or 551 Cancri b or c below the boiling point of water. The planet 55 Cancri d is 6 AU from its parent star, and it is cold out there; any water is likely to be frozen, unless there is an extreme greenhouse effect. One could imagine, of course, that this planet has a moon that is tidally heated, as Europa is, in which case liquid water could exist. Or perhaps, like the deep oceans on Earth, there are geothermal vents that bring heat up from the interior of the planet, melting the ice at least locally. Finally, HD 142415 has a semi-major axis close to that of Earth, and thus its equilibrium temperature should allow liquid water to exist. However, it is on a highly eccentric orbit and therefore alternately freezes and boils on a yearly cycle. Thus any life that has evolved on such a planet will have to deal with extreme temperature swings. One of the lessons of biology on Earth is that life has adapted to conditions we once would have thought completely inhospitable to life. Thus it is not inconceivable that creatures live in even this seemingly extremely difficult environment.

One final point. As already stated, all these planets are several times the mass of Jupiter, and like Jupiter, are probably gas giants (i.e., without a solid surface, and with a thick atmosphere mostly of hydrogen and helium). If we want to find an Earth-like surface with liquid oceans in which life might exist, perhaps we should be considering rocky or icy moons of these planets.

More recently, the Kepler spacecraft has observed the phenomenon of *planetary transits*, whereby an orbiting planet passes in front of its parent star, causing a brief dip in the brightness of that star. Kepler has discovered thousands of planets this way, including some with sizes and masses comparable to Earth. Such a planet is likely to be rocky, and some of these are at distances from their parent stars that liquid water can exist. We say that these are in the "habitable zone" or the

"Goldilocks zone" (not too hot, not too cold, but just right!). There is also a planet orbiting Proxima Centauri (the closest star to Earth, 4.2 light-years away), inferred from the radial velocity variations of its parent star. The planet has an inferred mass a bit higher than Earth. Its albedo and greenhouse effect are unknown, but it is possible that they are such as to yield a surface equilibrium temperature that allows liquid water to exist. However, it is so close to its parent star (an M star) that, like the Moon around Earth, it is tidally locked. If that is the case, then one face will always point to the star, making that face too hot, and the far side too cold to have liquid water.

66. Oceans in the solar system

66.a The mass m is the density ρ (Greek letter rho, often used to indicate density) times the volume V, and we know the density of water. We can calculate the volume from the formula given. We'll work to two significant figures:

$$m = \rho V = \frac{3}{4} 4\pi R^2 h \rho = 3\pi \left(6{,}371\,\text{km} \times 10^3\,\text{m/km}\right)^2 \times 3.5 \times 10^3\,\text{m} \times 1.0 \times 10^3\,\text{kg/m}^3$$

$$= 1.3 \times 10^{21}\,\text{kg}.$$

We did this using a calculator and then rounded to two significant figures. But we also checked the work by doing it approximately, to make sure we didn't type something wrong into the calculator. The above number can be written very roughly as:

$$3 \times 3 \times 6.4^2 \times 3 \times 10^{18}\,\text{kg}.$$

We know that $6.4^2 \approx 40$, and $3 \times 3 \approx 10$, so this is

$$3 \times 40 \times 10 \times 10^{18}\,\text{kg} = 1{,}200 \times 10^{18}\,\text{kg} = 1.2 \times 10^{21}\,\text{kg}.$$

Doing this kind of rough arithmetic got us pretty close; not perfect if we really need the second significant figure, but it is always a good idea to do the rough calculation as a check. Even when you are doing a precise calculation using a calculator, chances are, if the two differ significantly, you made a mistake in the precise version.

66.b We could calculate the mass of this ocean, and then take the ratio of that to the mass we calculated above. But let's take a somewhat different approach, which will be useful pedagogically (and lead to simpler arithmetic). Recognizing that the depth of the putative Martian ocean is poorly known, we are comfortable working to a single significant figure.

Our approach is to do as much algebra as possible before plugging in numbers:

$$\frac{m_{\text{Mars ocean}}}{m_{\text{Earth ocean}}} = \frac{4\pi r_{\text{Mars}}^2 h_{\text{Mars}}\rho}{3\pi r_{\text{Earth}}^2 h_{\text{Earth}}\rho} = \frac{4}{3}\left(\frac{r_{\text{Mars}}}{r_{\text{Earth}}}\right)^2 \frac{h_{\text{Mars}}}{h_{\text{Earth}}}.$$

Note that the expression for the Earth oceans includes the fact that they cover only 3/4 of Earth's surface. Now, the ratio of radii is roughly 1/2, while the ratio of ocean depths is about 1/3, so the ratio of the mass of the Martian and Earth oceans is about $\frac{4}{3}(1/2)^2 \times 1/3 \approx 1/9$, or 0.1. No calculator here, and we need not do it more carefully; given our uncertainty in the depth of the Mars ocean, we cannot express things any more precisely than this. But the conclusion is impressive; Mars may have once had 10% as much water as does the Earth now; that is a lot of water!

Where did that water go? Some of it may be tied up in subsurface water-rich mineral compounds or in extensive underground water/ice fields. Liquid water cannot last for long on Mars' surface now; the atmosphere is so thin that any water would quickly vaporize.

66.c The ratio of the masses of the two oceans is just the ratio of their volumes (after all, they have the same density). The volume is given by their surface areas times their depths (times the 3/4 factor for Earth):

$$\frac{m_{\text{Europa ocean}}}{m_{\text{Earth ocean}}} = \left(\frac{r_{\text{Europa}}}{r_{\text{Earth}}}\right)^2 \frac{h_{\text{Europa}}}{\frac{3}{4} h_{\text{Earth}}} = \left(\frac{1,460}{6,371}\right)^2 \times \frac{4}{3} \times \frac{100}{3.5} \approx 2.$$

Again, we're allowed to compute this to a single significant figure, as we only know the depth of the Europan ocean to one significant figure. There is twice as much water on Europa as on Earth.

It is hypothesized that the heat to keep this water in a liquid state comes from Europa itself. As the moon moves through its elliptical orbit, it is kneaded by the changing tidal forces from Jupiter, and that squeezing causes internal friction, which generates heat. Given this extensive water reservoir and a source of heat, it is hypothesized that life may have evolved on Europa, perhaps swimming beneath the thick ice covering the moon. Astronomers are eager to send spacecraft to Europa to take a look, but it would be a challenge, to put it mildly, to drill through the many kilometers of ice at the surface to get to the liquid water.

67. Could photosynthetic life survive in Europa's ocean?

67.a The solar flux is

$$b = \frac{L}{4\pi d^2} = \frac{4 \times 10^{26} \text{ watts}}{12 \times (5.2 \times 1.5 \times 10^{11} \text{ meters})^2} \approx 50 \text{ watt/m}^2.$$

This is *much* more than the minimum flux levels for photosynthesis, so yes, such organisms could easily photosynthesize at Europa's distance from the Sun.

67.b The total area of liquid water exposed to space is 10^{-7} of the full surface area of the planet, $4\pi R^2$, or

$$\text{Area} = 10^{-7} \times 4\pi (1600 \text{ km})^2 \approx 3 \text{ km}^2.$$

Each square kilometer is 10^6 square meters, so the total power available to these hypothetical microorganisms is the product of this area and the solar flux calculated in part **a**, divided by two because only half of Europa is illuminated at any one time:

$$\text{Power} = \frac{1}{2} \times 3 \times 10^6 \text{ m}^2 \times 50 \text{ watt/m}^2 = 8 \times 10^7 \text{ watt}.$$

67.c One year is 3×10^7 seconds, and remembering that 1 watt is 1 Joule per second, the total energy received in the Europan cracks over the course of a year is

$$8 \times 10^7 \text{ Joules/sec} \times 3 \times 10^7 \text{ sec} = 2.4 \times 10^{15} \text{ Joules}.$$

The total carbon biomass produced in that time is just the above times the amount of biomass produced per Joule:

$$2.4 \times 10^{15} \text{ Joules} \times 2 \times 10^{-10} \text{ kg/Joule} = 5 \times 10^5 \text{ kg}.$$

Wow, 500 tons; that's a lot! And at 10^{-15} kg per microorganism, that's an astonishing 5×10^{20} organisms, an impressive number!

67.d This is straightforward: if we have 500,000 kg of biomass in 1 year, then we get 1,000 times that, or 5×10^8 kg in 1,000 years. That is is still tiny compared with the biomass of Earth (the ratio is $\frac{5 \times 10^8 \, \text{kg}}{10^{15} \, \text{kg}} = 5 \times 10^{-7}$), but this number is suggestively large enough to make one want to go looking. While the ice crust of Europa is thought to be 10 kilometers thick, making it challenging to probe the liquid water underneath, we've seen that the crust does exhibit cracks, where it is thought that liquid water may occasionally come to the surface. This would be the place to send a visiting spacecraft to look for signs of life.

68. An essay on liquid water

Water is liquid over a range of temperatures; for normal sea-level pressure, this range is 0–100°C (273–373 K).

There is strong evidence that Mars once had extensive liquid water on its surface, including (but not limited to!):

- (Now dry) riverbeds and deltas,
- Mineral deposits detected by the Mars rovers that must have been laid down in water,
- Ice deposits visible on the surface, and
- Gullies thought to have been cut by water.

Thus Mars probably once was warm enough (via an atmosphere that gave rise to a greenhouse effect) to allow liquid water to exist. Standing liquid water cannot currently exist on Mars: it is both too cold (so the water would freeze) and the atmospheric pressure is too low (so the water would boil). However, liquid water could possibly exist beneath the surface of the planet.

Europa, a moon of Jupiter, is thought to have a liquid ocean perhaps 90 kilometers thick under a top layer of ice, which is heated both by tidal forces from Jupiter and radioactive decay from the moon's core; cracks in the surface ice imply that the ice floats on water. Water geysers have been observed on Enceladus, a moon of Saturn. There are other bodies in the solar system that are known to be dry, including Mercury, Venus, and the Moon (although there is strong circumstantial evidence for water ice in deep craters on the Moon that are permanently in shadow). Comets are largely made of water ice, and it has been hypothesized that much of Earth's water came from comets; indeed, water on early Earth may have vaporized completely due to asteroid impacts.

Planets (including Earth) are heated by the Sun. A planet's temperature is determined by an equilibrium between the sunlight absorbed and blackbody radiation that the planet gives off. Planets don't absorb all the light that falls on them; the albedo A expresses the fraction of sunlight that is reflected. Similarly, the atmosphere traps some of the infrared radiation given off by Earth (the greenhouse effect), raising the temperature somewhat higher than it would be without an atmosphere. The equation for the equilibrium temperature of a planet of albedo A a distance d from a star of surface temperature T and radius R is

$$T_{\text{planet}} = T(1 - A)^{\frac{1}{4}} \sqrt{\frac{R}{2d}}.$$

Ignoring the greenhouse effect for Earth gives a temperature well below the freezing point of water. However, including the albedo of Earth and the greenhouse effect gives a temperature of 300 K, very close to the right value. The interior of Earth is hot, but the surface of the planet receives little direct heating from its interior via geothermal effects.

The fact that liquid water has remained on the Earth's surface for billions of years, even though the Sun was less luminous in the past, suggests that the greenhouse effect was greater in the past than it is now. The principal greenhouse gases on Earth are water vapor, carbon dioxide, and methane. There is a complex carbon cycle with a negative feedback on Earth, which regulates the temperature. It works as follows: CO_2 dissolves in rain in Earth's atmosphere, or directly in the oceans themselves; it then solidifies as calcium carbonate ($CaCO_3$) via biological effects (creatures making shells) or abiotic effects. The calcium itself was dissolved out of rocks by carbon-dioxide-rich rain. This now-solid material falls to the ocean floors, and eventually gets subducted into Earth's mantle by plate tectonics. It is released again in the form of volcanism, returning CO_2 to the Earth's atmosphere. This cycle operates on timescales of a few hundred million years. If Earth becomes hotter for some reason, there is more rain, and more CO_2 gets washed out of the Earth's atmosphere, decreasing the greenhouse effect and decreasing the temperature. On Mars, plate tectonics seems to have come to a halt, and this feedback doesn't happen: as a consequence, there is essentially no greenhouse effect, and the surface is cold. In contrast, on Venus, whatever water might have once existed has been disassociated (i.e., torn apart into its component atoms) by ultraviolet light from the Sun (the ozone in Earth's atmosphere blocks this UV radiation, preventing the disassociation of water). On Venus, the escape speed is smaller than the thermal speed of hydrogen, meaning that the disassociated hydrogen is lost from the planet forever. The atmosphere therefore has no water, and so there is no mechanism to take CO_2 out of the atmosphere; the atmosphere is loaded with CO_2. Thus Venus has an extreme greenhouse effect, and its surface is beastly hot.

69. How many stars are there?

69.a We are given the dimensions of the Milky Way and are told the typical distance between stars. The way to think about this is to ask how many stars we could pack into the entire volume of the Milky Way. What counts here is the volume that each star, plus the empty volume around it, occupies. How should we think about this? If the stars are 4 light-years apart, then we can think of each sitting at the center of a cube of side 4 light-years (draw a picture to convince yourself that this indeed puts the stars 4 light-years apart). That is, there is one star in a cube of volume $(4 \, \text{ly})^3$. The volume of the Milky Way itself is its thickness times the area of the disk, namely, $\pi R^2 l$, where $R = 50,000$ light-years is the radius of the disk, and $l = 1,200$ light-years is its thickness. Thus the number of stars we can put into that volume is that volume divided by the volume each star occupies:

$$\text{Number of stars} = \frac{\pi R^2 l}{(4 \, \text{ly})^3} \approx \frac{3 \times (5 \times 10^4 \, \text{ly})^2 \times 1.2 \times 10^3 \, \text{ly}}{(4 \, \text{ly})^3} \approx 1.5 \times 10^{11}.$$

That is, there are roughly 150 billion stars in the Milky Way. This is only a factor of 2 different from the modern estimate of about 300 billion stars.

Note that all distances above were expressed in light-years, so the units of the numerator and denominator (both $(\text{light-year})^3$) canceled each other. If that hadn't been the case, we would have needed to convert the various units to be consistent.

69.b This is a very similar problem to part **a**. Now we're packing galaxies into the universe as a whole. By exactly the same reasoning as in part **a**, we can envision each galaxy occupying a volume of a cube $r = 15$ million light-years on a side. We're putting this into a sphere of radius $R = 45$ billion light-years:

$$\text{Number of galaxies} = \frac{\frac{4}{3}\pi R^3}{r^3} \approx 4\left(\frac{R}{r}\right)^3 = 4 \times \left(\frac{45 \times 10^9 \,\text{ly}}{15 \times 10^6 \,\text{ly}}\right)^3 = 4 \times (3,000)^3 \approx 1 \times 10^{11}.$$

To a good approximation, there are about 100 billion galaxies in the visible universe, a number in the same ballpark as the number of stars in the Milky Way Galaxy. Note the advantages of doing the algebra before plugging in numbers. In particular, life gets much easier when we write $(R/r)^3$ rather than R^3/r^3. The two expressions are of course equivalent, but it is much easier on your brain to do the former once you plug in the numbers.

If the universe is about 14 billion years old, why do we give an extent of 45 billion light-years, not 14 billion light-years? The light from the most distant galaxies indeed has been traveling to us for 14 billion years, but in that time, the universe has continued to expand; the distance to those galaxies now is even larger.

It is worth remarking that our calculation is based on the numbers of big luminous galaxies. In fact, galaxies have a rather continuous distribution of luminosities, and there are many more low-luminosity galaxies than high-luminosity galaxies like the Milky Way. Recent estimates for the total number of all galaxies in the observable universe, including the low-luminosity galaxies as well, are higher still, about 2 trillion galaxies. But most of the stars in the universe are found in the big galaxies like our own.

69.c The total number of stars in the visible universe is just the product of the total number of galaxies and the number of stars per galaxy:

$$\text{Number of stars in the universe} = 1 \times 10^{11} \text{ galaxies} \times 1.5 \times 10^{11} \text{ stars/galaxy}$$

$$= 1.5 \times 10^{22} \text{ stars}.$$

That is indeed a seriously big number: 15 sextillion.

The terms for big numbers in English are very similar to each other: "million," "billion," "trillion," and so on, and so often we interpret them as all roughly equivalent to "something really large." It is important to remember, however, just how different they are from one another: even though a million is absolutely tiny compared to a trillion, they are often mixed up in the popular press. Sticking with scientific notation is a way to avoid confusion.

70. The distance between stars
We are asked for the ratio of 4 light-years to 0.01 AU. Let's just do it by converting both distances to kilometers:

$$\frac{4\,\text{ly}}{0.01\,\text{AU}} = \frac{4 \times 10^{13}\,\text{km}}{1.4 \times 10^6\,\text{km}} \approx 3 \times 10^7.$$

This ratio is thus a factor of 30 million! Space is *really* empty.

So if we scale the Sun down to the size of a basketball, the next star is 10 inches $\times\, 3 \times 10^7 = 3 \times 10^8$ inches away. There are roughly 40 inches in a meter, and thus 40,000 inches in a kilometer, so the nearest stars are about

$$\frac{3 \times 10^8 \, \text{inches}}{4 \times 10^4 \, \text{inches/km}} \approx 7000 \, \text{km}$$

away, roughly the distance from New York to São Paulo, Brazil.

71. The emptiness of space

71.a The density is the mass divided by the volume. For this problem, we could take the total number of stars in the Milky Way, and given a rough estimate of its dimensions (see, for example, problem 69), we could calculate a volume. Dividing the number of stars times the mass of each by the volume would give us a density. But we could just as well do this calculation using any representative chunk of the Milky Way. In particular, let us examine a single star and the volume it occupies. If the stars are 4 light-years apart, we can think of each star as sitting at the center of its own cube, with sides 4 light-years long. Thus the density ρ is the mass of the one star, divided by the volume of the cube:

$$\rho = \frac{\text{Mass}}{\text{Volume}} = \frac{2 \times 10^{33} \, \text{g}}{(4 \, \text{ly} \times 10^{13} \, \text{km/ly} \times 10^5 \, \text{cm/km})^3}$$
$$= \frac{2}{64} \times 10^{-21} \, \text{g/cm}^3 \approx 3 \times 10^{-23} \, \text{g/cm}^3.$$

For comparison, the density of water is 1 gram per cubic centimeter. This density we've just calculated is 300 times lower than the most extreme laboratory vacuum. Space is incredibly empty indeed!

We'll see in the next part that the Milky Way is actually quite overdense relative to the universe as a whole.

71.b In part **a**, we could calculate the density of the whole galaxy by dividing the mass of a single star by the volume it occupied (a cube of side 4 light-years). We're going to use the same approach here: the average density of the universe ρ is the mass of a single galaxy divided by the volume that it occupies—namely, a cube of side 15 million light-years:

$$\rho = 1.5 \times 10^{11} \, \frac{\text{stars}}{\text{galaxy}} \times 2 \times 10^{33} \, \frac{\text{g}}{\text{star}} \times \frac{1 \, \text{galaxy}}{(1.5 \times 10^7 \, \text{ly} \times 10^{13} \, \text{km/ly} \times 10^5 \text{cm/km})^3}.$$

OK, this is going to keep us busy. In the denominator, $1.5^3 \approx 3$ to a single significant figure, and $1.5 \times 2 = 3$ in the numerator. So those terms cancel, and we're just left with factors of 10:

$$\rho \approx 10^{11+33-3(7+13+5)} \, \text{g/cm}^3 = 10^{-31} \, \text{g/cm}^3.$$

That is an impressively small number indeed, 300 million times smaller than the already incredibly low density that we calculated previously for an individual galaxy. This reflects the large amount of empty volume between galaxies. To put it another way, galaxies are overdense by a factor of 300 million relative to the universe as a whole!

We've ignored the dark matter associated with galaxies in this calculation. Including dark matter in the Milky Way raises its mass to about a trillion solar masses, but this does not yet include dark matter associated with lower-mass galaxies. The best current estimate for the total mass density of the universe, including dark matter and dark energy, is about a factor of 100 larger than what we have calculated, or about 10^{-29} grams per cubic centimeter.

Let's develop some intuition about how small the density we calculated above really is by calculating the radius of a sphere of that density with a mass equal to a single hydrogen atom, and that of a sphere of the same density containing the mass of a human being.

A sphere of density ρ with radius r has a mass $\frac{4\pi}{3}r^3\rho$. We have to solve for r in two cases: a volume containing the mass of a single hydrogen atom, and one containing the mass of a person. Let's just do it, approximating π by 3. First for the mass of an atom:

$$4r^3 \times 10^{-31}\,\frac{\text{g}}{\text{cm}^3} = \frac{1}{6} \times 10^{-23}\,\text{g}.$$

Note the way we have rewritten the mass of the atom; you may recognize Avogadro's number in there.

$$r^3 = \frac{1}{24} \times 10^{-23+31}\,\text{cm}^3 \approx 4 \times 10^6\,\text{cm}^3.$$

Can we do cube roots without a calculator? Yes, if we are happy with an answer to a single significant figure. The cube root of 4 is between 1.5 and 2; let's round up to 2. And the cube root of 10^6 is just 100. So we find that the radius of that sphere is about 200 centimeters, or 2 meters. That's a big volume to hold only a single atom!

OK, let's redo this for a mass of 80 kilograms, or 8×10^4 grams. The setup is the same:

$$4r^3 \times 10^{-31}\,\frac{\text{g}}{\text{cm}^3} = 8 \times 10^4\,\text{g}$$

$$r^3 = 2 \times 10^{35}\,\text{cm}^3.$$

We'll rewrite this as $200 \times 10^{33}\,\text{cm}^3$. The cube root of 200 is about 6, so our final answer is

$$r = 6 \times 10^{11}\,\text{cm}.$$

Remember that there are 10^5 cm in a kilometer, so the radius of a sphere at the density we've calculated, with a mass equal to that of a human being, is 6×10^6 km, almost five times the diameter of the Sun!

72. Squeezing the Milky Way

We can do this problem approximately as follows. Let's take each star to be as large as the Sun, with a radius of $r_{\text{star}} = 700,000$ kilometers. This is a crude approximation; less massive and therefore smaller stars are somewhat more common than Sun-like stars, but let's ignore such details for the moment. We're putting all the stars into a ball that will have the same density as that of a single star. The density is the mass divided by the volume, so we state:

$$\text{Density of single star} = \text{Density of ball of stars},$$

$$\frac{\text{Mass of single star}}{\text{Volume of single star}} = \frac{\text{Mass of ball of stars}}{\text{Volume of ball of stars}},$$

$$\frac{M_{\text{star}}}{\frac{4}{3}\pi r_{\text{star}}^3} = \frac{1.5 \times 10^{11} M_{\text{star}}}{\frac{4}{3}\pi r_{\text{ball}}^3}.$$

Solving this equation for the radius of the ball of stars gives us

$$r_{\text{ball}} = r_{\text{star}} \times \left(1.5 \times 10^{11}\right)^{1/3}.$$

We can do this approximately, even without a calculator, by writing $1.5 \times 10^{11} = 150 \times 10^9$. We recognize that 150 is just a bit more than 125, which is 5^3, so to a decent approximation:

$$r_{\text{ball}} = 700{,}000\,\text{km} \times 5 \times 10^3 = 3.5 \times 10^9\,\text{km}.$$

There are 1.5×10^8 kilometers in an AU, so this comes out to about 23 AU. That is, if we were to squeeze all the stars in the Milky Way into a ball in which the stars were just touching, it wouldn't reach even out to the orbit of Neptune around the Sun. This reminds us again how large the space between the stars is relative to their sizes.

Note that we did not worry about the small gaps left when stacking spheres together. The error made in ignoring these gaps was much smaller than uncertainties in the precise numbers of stars in the Milky Way, or the sizes of each of them.

There's another subtlety to this problem. The radius of the black hole of a given mass (the *Schwarzschild radius*) is proportional to that mass. The Schwarzschild radius of the Sun is about 3 kilometers, so the Schwarzschild radius of 1.5×10^{11} solar masses is therefore

$$3\,\text{km} \times 1.5 \times 10^{11} \times \frac{1\,\text{AU}}{1.5 \times 10^8\,\text{km}} = 3000\,\text{AU}.$$

That is, this ball of stars is smaller than the Schwarzschild radius of the black hole of that mass. Which is to say, if we were somehow to bring this group of stars together, it would instantly collapse into a black hole. How cool is that?

73. A star is born

We are asking what volume of interstellar space encloses a mass of gas and dust comparable to the mass of a star. The mass of a volume V is the product of that volume and the density, so we can write:

$$\rho V = M; \qquad V = \frac{M}{\rho}.$$

We know both the mass (1 solar mass) and the density, so we can calculate:

$$V = \frac{2 \times 10^{30}\,\text{kg}}{3 \times 10^{-20}\,\text{kg/m}^3} = \frac{2}{3} \times 10^{50}\,\text{meter}^3.$$

To understand how large a volume that is, let us determine what the radius of a sphere of this volume is:

$$\frac{4}{3}\pi r^3 = \frac{2}{3} \times 10^{50}\,\text{m}^3,$$

so

$$r = \left(\frac{10^{51}}{60}\right)^{1/3}\,\text{meters},$$

where we approximated $2\pi = 6$, and rewrote the fraction in a way that makes the arithmetic we're about to do easy. In particular, note that the cube root of 10^{51} is 10^{17}. Moreover, 60 is close to $64 = 4^3$. So the cube root isn't difficult:

$$r \approx \frac{10^{17}}{4}\,\text{meters} = 2.5 \times 10^{16}\,\text{meters}.$$

Remember that there are about 10^{16} meters in a light-year, so this is 2 or 3 light-years. A star like the Sun was formed from gas that may have originally been spread over a volume several light-years across.

In fact, stars typically form not out of gas at the mean density of the interstellar medium, but out of relatively dense molecular clouds, so-called because the gas in them is mostly in molecular form (H_2, CO, and H_2O are among the most common molecules).

74. A massive black hole in the center of the Milky Way

74.a This is a straightforward application of the small-angle formula. In this case, we are asking what length s subtends $\theta = 0.1$ arcsec at a distance d of the Milky Way Center (i.e., 8 kiloparsecs). The small-angle formula tells us that this distance is

$$s = d\theta = 8,000\,\text{pc} \times 0.1\,\text{arcsec}.$$

At this point, you are probably wondering when to convert from arcseconds to radians. But remember, from the definition of parsecs, the small-angle formula works for θ in arcseconds, d in parsec, and s in AU. So if we do the arithmetic above as-is, we'll get an answer in AU. The answer is that the semi-major axis s is 800 AU, or, with 1.5×10^8 km/AU, 1.2×10^{11} km.

74.b Remember the formula relating the speed of an orbit v, its radius r, and the mass of the central object M:

$$v^2 = \frac{GM}{r}.$$

However, in this case, we have the period P of the orbit; let's rewrite the equation by remembering that $v = \frac{2\pi r}{P}$:

$$\left(\frac{2\pi r}{P}\right)^2 = \frac{GM}{r}.$$

Solving for M, we find:

$$M = \frac{4\pi^2 r^3}{P^2 G}.$$

We've just rederived Newton's form of Kepler's third law. Now, we could plug in all the numbers in the usual way, converting everything to consistent units. But an easier way to proceed is to remember that we can write down exactly the same equation for the orbit of Earth around the Sun (where $r = 1$ AU and $P = 1$ year). If we then take the ratio of the two equations, lots of things cancel, the units are easy, and we find the much simpler form

$$\frac{M_{\text{MW center}}}{M_{\text{Sun}}} = \left(\frac{800\,\text{AU}}{1\,\text{AU}}\right)^3 \times \left(\frac{1\,\text{year}}{11.5\,\text{year}}\right)^2.$$

Let's see. We have $800^3 \approx 5 \times 10^8$, and $11.5^2 \approx 130$, so

$$\frac{M_{\text{MW center}}}{M_{\text{Sun}}} = \frac{5 \times 10^8}{130} \approx 4 \times 10^6.$$

Note that we automatically have a result in the units we wish, namely, solar masses. We have found that the central mass around which the star is orbiting has a mass 4 million times that of the Sun. Wow!

74.c The density is the mass divided by the volume. Assuming the object is spherical, its largest possible volume is that of a sphere of radius 45 AU. And its mass is what we've just calculated in part **b**. Thus the smallest possible density (i.e., the lower limit) is the ratio of this mass to the largest possible volume. Let's calculate. There are no clever tricks to save us time; we want results in grams per cubic centimeter, so we just have to convert:

$$\rho = \frac{M}{V} = \frac{4 \times 10^6 \, M_{\text{Sun}} \times 4 \times 10^{33} \, \text{g}/M_{\text{Sun}}}{\frac{4}{3}\pi \left(45 \, \text{AU} \times 1.5 \times 10^{13} \, \text{cm}/\text{AU}\right)^3}.$$

We'll do the usual approximation that $\pi = 3$. The quantity $1.5^3 \approx 4$, and we'll write $45 \approx 100/2$, so lots of numbers cancel and we get

$$\rho \approx 8 \times 10^{6+33-45} \, \frac{\text{g}}{\text{cm}^3} \approx 10^{-5} \, \frac{\text{g}}{\text{cm}^3}.$$

While this density is not terribly high, any ordinary group of stars of this total mass packed into such a small volume would shine incredibly brightly and would be immediately obvious to our telescopes. The only known explanation for such a large and invisible mass in such a small volume is a black hole.

74.d A star in an orbit of radius 100 parsecs is subject to the gravitational pull of all the material in a sphere inside its orbit. That material includes both the black hole and all the stars in that sphere. We've calculated the former. For the latter, we're talking about 1000 stars per cubic parsec over a volume of a sphere of radius 100 parsecs and thus a total of

$$1{,}000 \, \frac{\text{stars}}{\text{pc}^3} \times \frac{4}{3}\pi (100 \, \text{pc})^3 \approx 4 \times 10^{3+6} = 4 \times 10^9 \, \text{stars}.$$

Or, assuming that each star has a mass similar to that of the Sun, 4 billion solar masses. The mass of the black hole is only $1/1{,}000$ of that, essentially negligible. The point of this exercise is that if we want to measure the mass of the black hole, we can do so only very close in, where its mass dominates; out at 100 parsecs, the mass is dominated by the stars interior to that radius.

The black hole in the center of the Milky Way is far from unique. Indeed, as far as we can tell, such supermassive black holes reside in the center of essentially all large galaxies. The Milky Way black hole is close enough that we can resolve the orbits of individual stars to infer a mass. For other galaxies, we work instead with the combined light of multiple stars and use the Doppler shift to infer their average speeds. But the conclusion is the same: large black holes reside at the centers of these galaxies as well.

75. Supernovae and the Galaxy

75.a Two percent of the mass of the Galaxy is $2 \times 10^9 \, M_{\text{Sun}}$. If each supernova contributes 1 solar mass of heavy elements, we're talking about 2×10^9 supernovae.

75.b Over the 1.4×10^{10} years since the Big Bang, if the rate has been constant, we expect 1.4×10^8 supernovae to have gone off. The actual number needed is about ten times that, so the supernova rate should have been appreciably higher in the past. Indeed, we come to a similar conclusion for most galaxies in the nearby universe. Given that supernovae represent the explosion of massive, and thus relatively recently formed stars, we would predict that galaxies used to form stars at a much more rapid clip than they do today. Luckily, we can observe that directly: because of the finite speed of light, we are seeing distant galaxies as they were in past, and it is true that the rate of star formation in those distant galaxies is, on average, much higher than that for nearby galaxies.

76. Dark matter halos

76.a Consider a star of mass m at radius r going around a circle at speed v. The acceleration that gives rise to the circular motion is $a = v^2/r$. This is due to a force, namely the gravitational attraction of the material within radius r. This material acts as if it were all at the center (i.e., a distance r away). Indicating the mass of the material in the radius r as $M(<r)$, we find

$$F = ma = m\frac{v^2}{r} = \frac{GM(<r)m}{r^2}.$$

We are asked for the mass of the Milky Way within r, so solving for $M(<r)$ gives

$$M(<r) = \frac{v^2 r}{G}.$$

Note that the result does not depend on the mass of the star m with which we started. This often arises in gravitational problems: the acceleration of all objects at a given point in a gravitational field is the same. Note also that for a flat rotation curve (as we observe for the Milky Way), the mass grows in proportion to the radius. This is a consequence of the fact that the Milky Way is embedded in a massive dark matter halo; as one goes farther and farther out, one feels the gravity from an ever-increasing amount of mass. We will explore the consequences of this in what follows.

76.b The average density within radius r is the mass divided by the volume. We consider a sphere of radius r; using the mass calculated in part **a**, we find

$$\text{Density} = \frac{M(<r)}{\text{Volume}} = \frac{v^2 r/G}{4\pi r^3/3} = \frac{3v^2}{4\pi G r^2}.$$

So even though the mass increases as you go farther out, the volume increases faster, and the density actually decreases. The density of dark matter drops off as one moves from the center of the Milky Way.

76.c Determining the mass can be done in two ways. First, we simply use the equation for $M(<r)$ in the form we found in part **a**, using as an upper limit a radius of half the distance between galaxies, namely, 1 million light-years:

$$M_{\text{galaxy}} < \frac{v_{\text{galaxy}}^2 r_{\text{galaxy}}}{G} = \frac{\left(2.2 \times 10^5\,\text{m/sec}\right)^2 \times 10^6\,\text{ly} \times \frac{10^{16}\,\text{m}}{1\,\text{ly}}}{\frac{2}{3} \times 10^{-10}\,\text{m}^3\,\text{sec}^{-2}\,\text{kg}^{-1}}$$

$$= 5 \times \frac{3}{2} \times 10^{10+6+16+10}\,\text{kg} = 7 \times 10^{42}\,\text{kg}.$$

The $<$ ("less than") sign above indicates that this is an upper limit, assuming the Milky Way's dark matter halo really extends to 1 million light-years.

One solar mass is 2×10^{30} kg, so we get for a final answer a mass of $3.5 \times 10^{12}\,M_{\text{Sun}}$, or 3.5 trillion solar masses. The vast majority of this is dark matter.

Another way to do this calculation is by scaling from the solar system: Earth is going around the Sun at 30 km/sec at a distance of 1 AU. We can use the same formula to infer the mass of the Sun:

$$M_{\text{Sun}} = \frac{v_{\text{Earth}}^2 r_{\text{Earth}}}{G}.$$

Taking the ratio of this expression to the one we have for the Milky Way as a whole yields

$$\frac{M_{\text{galaxy}}}{M_{\text{Sun}}} < \left(\frac{v_{\text{galaxy}}}{v_{\text{Earth}}}\right)^2 \frac{r_{\text{galaxy}}}{r_{\text{Earth}}} = \left(\frac{220}{30}\right)^2 \frac{10^6\,\text{ly} \times 10^{16}\,\text{m/ly}}{1.5 \times 10^{11}\,\text{m}} = \frac{7^2}{1.5} \times 10^{11}.$$

Thus $M_{\text{galaxy}} = 3 \times 10^{12}\,M_{\text{Sun}}$. The only reason we didn't get exactly the same answer as above was because of rounding errors.

In fact, this value is an overestimate of the true mass of the Milky Way; the assumption that the dark matter halo reaches halfway to the Andromeda galaxy is not correct. There is another way to estimate the Milky Way mass, recognizing that the Milky Way, the Andromeda galaxy, and a number of smaller companion galaxies are all orbiting under their mutual gravitational influence. Measuring their motions, one infers a total Milky Way mass of roughly 1×10^{12} solar masses, a factor of 3 lower. We may thus conclude that the dark matter halo of our Galaxy extends to about 1/3 of the assumed 1 million light-years, or about 300,000 light-years.

77. Orbiting galaxy

77.a We are given an angular velocity (1×10^{-3} arcseconds per year), and we want to know the actual velocity, given that the LMC is 1.5×10^5 light-years away. This is a problem in unit conversion and the small-angle formula.

$$1 \times 10^{-3}\frac{\text{arcsec}}{\text{year}} \times \frac{1\,\text{radian}}{2 \times 10^5\,\text{arcsec}} \times 1.5 \times 10^5\,\text{ly} \times \frac{10^{13}\,\text{km}}{1\,\text{ly}} = 7.5 \times 10^9\,\frac{\text{km}}{\text{year}}.$$

Converting to km/sec, we find

$$7.5 \times 10^9\,\frac{\text{km}}{\text{year}} \times \frac{1\,\text{year}}{3 \times 10^7\,\text{sec}} = 250\,\text{km/sec}.$$

Thus the LMC is orbiting the Milky Way at a speed comparable to the orbital speed of the Sun (220 kilometers per second), even though it is much farther away. As is described in *Welcome to the Universe,* the Milky Way (and most spiral galaxies) exhibit flat rotation curves, in which the rotation speed is almost independent of distance from the center, a reflection of the existence of a massive halo of dark matter. This shows that the Milky Way halo extends to a radius of at least 150,000 light-years.

77.b The easy way to do this is to realize that one orbit is $360° = 360 \times 3{,}600$ arcsec. Thus to go all the way around takes

$$\frac{360 \times 3600\,\text{arcsec}}{1 \times 10^{-3}\,\text{arcsec/year}} = 1.3 \times 10^9\,\text{years}.$$

Another way to do this is to calculate the circumference of the circle and divide by the speed. This is much more work, but it will get you the right answer.

Note that this orbital time is much longer than the time (250 million years) it takes for the Sun to make one orbit around the Milky Way. The LMC is orbiting at about the same speed but has much further to travel to make one orbit.

77.c The easiest way to do this is to scale directly from the equivalent result at 25,000 light-years, the radius of the Sun's orbit. The LMC's orbital radius is six times larger, and the period is five times longer. The mass scales as a^3/P^2 by Kepler's third law, which gives a factor of $6^3/5^2 = 8$. Thus the mass out to the LMC is eight times that at the solar circle, or $8 \times 10^{11}\ M_{\mathrm{Sun}}$.

Note that this problem could be done using Newton's form of Kepler's third law, but that requires much more work.

It is tempting to do this problem a different way, by assuming that the density of the Milky Way is constant, and thus the mass is proportional to the volume. This is not correct; the density of the Milky Way drops off dramatically with distance from the center (see problem 76.**b**).

78. Detecting dark matter

78.a We know how to calculate the mass in the shell:

$$M(<(r+\Delta r)) - M(<r) = \frac{v^2(r+\Delta r)}{G} - \frac{v^2 r}{G} = \frac{v^2 \Delta r}{G}.$$

The volume of the shell is its surface area, $4\pi r^2$, times its thickness Δr. Thus the density ρ is the mass divided by this volume:

$$\rho = \frac{v^2 \Delta r}{G} \frac{1}{4\pi r^2 \Delta r} = \frac{v^2}{4\pi r^2 G}.$$

The thickness of the shell has dropped out of the equation.

78.b Let us plug in the numbers, converting throughout to MKS units:

$$\rho = \frac{v^2}{4\pi r^2 G} = \frac{(220\,\mathrm{km/sec} \times 10^3\,\mathrm{m/km})^2}{4\pi(25{,}000\,\mathrm{ly} \times 10^{16}\,\mathrm{m/ly})^2 \times \frac{2}{3} \times 10^{-10}\,\mathrm{m^3/sec^2\,kg}}.$$

Let us make the approximation that $\pi = 3$ and that $22 = 25$; there is also a factor of 4×2 in the denominator that we approximate as 10. This simplifies things, and we find, after gathering all the powers of 10:

$$\rho - 10^{-21}\,\mathrm{kg/m^3}.$$

We were asked to express this in terms of grams per cubic centimeter. There are 1,000 grams in a kilogram. And with 100 centimeters per meter, there are 10^6 cubic centimeters in a cubic meter. Thus the density is $10^{-24}\,\mathrm{g/cm^3}$. That's an impressively small density!

78.c The mass of a proton in grams, to a very good approximation, is $\frac{1}{6} \times 10^{-23}$ grams. Einstein tells us that this is equivalent to an energy of mc^2, or

$$E = \frac{1}{6} \times 10^{-23}\,\mathrm{g} \times (3 \times 10^{10}\,\mathrm{cm/sec})^2 = \frac{3}{2} \times 10^{-3}\,\mathrm{erg}.$$

Converting to electron volts means dividing by $1.6 \times 10^{-12} = \frac{1}{6} \times 10^{-11}$ eV/erg, which gives

$$E = 9 \times 10^8\,\mathrm{eV},$$

or 0.9 GeV (where a giga-electron volt, GeV, is 10^9 eV). If we do this with a calculator, to another significant figure, we get 0.94 GeV, or 940 MeV. It is quite standard in physics to quote the masses (or, strictly speaking, the mass-energies) of elementary particles in electron volts. For example, the electron has a mass of 511 keV (511 thousand electron volts), and the newly discovered Higgs boson has a mass of 125 GeV.

The number density of dark matter particles is their mass density (which we calculated above, after we take into account the factor of half that is the fraction of dark matter), divided by the mass of each one. We have their mass density in grams, so we'll work in grams here. A particle with a mass equal to 1,000 times that of a proton has a mass of $\frac{1}{6} \times 10^{-20}$ grams. So the number density is

$$n = \frac{\rho}{m} = \frac{0.5 \times 10^{-24}\,\text{g/cm}^3}{\frac{1}{6} \times 10^{-20}\,\text{g/particle}} = 3 \times 10^{-4}\,\text{particles/cm}^3.$$

We are embedded in this sea of dark matter particles. But they rarely interact with ordinary matter, so they are difficult to detect!

78.d In a time t, the WIMP wind moves a distance vt. Thus consider the wind hitting one face of the cube with area L^2; this wind will sweep through the cube. In time t, it will sweep out a volume $vt \times L^2$, and because we know the number density of WIMPs (n), the total number of WIMPs passing that face of the cube is the product of the two factors: vtL^2n. We want the number per unit time, so we simply divide by the time to get the result vL^2n. You should confirm that this has the units of number of WIMP particles per unit time.

78.e Consider a long skinny circular cylinder with base of area σ and height L (that's the cross-section of the particle, swept out through the detector). It has volume σL. Now, the number density of nucleons in the detector is the mass density of nucleons divided by the mass of each one. The mass density is the mass of the detector, M, divided by the volume L^3, so the number density of nucleons is $\frac{M}{L^3 m}$. Thus the probability that it interacts with one of the nucleons is the product of these two quantities: $\frac{M\sigma}{L^2 m}$.

78.f An interaction happens when a WIMP's cross-section sweeps out a volume that includes a nucleon in the detector. We've just calculated the probability of an interaction per WIMP passing through the detector, and this, multiplied by the rate at which WIMPs pass through the detector, is the total rate of interactions. Multiplying these two together then gives

$$vL^2n \times \frac{M\sigma}{L^2 m} = \frac{nv\sigma M}{m}.$$

Note that the dimensions of the box, L, have dropped out. You should convince yourself that the rate derived has units of $1/\text{time}$, as it should; it is the number of interactions per unit time.

78.g We have all the numbers we need. The speed $v = 220$ km/sec is the rotational speed of the Milky Way; $m =$ the mass of a proton $= 1.6 \times 10^{-24}$ g. We calculated n in part **c**. Plugging this all in, we find

$$\frac{nv\sigma M}{m} = \frac{3 \times 10^{-4}\,\text{cm}^{-3} \times 2.2 \times 10^7\,\text{cm/sec} \times 131^2 \times 10^{-45}\,\text{cm}^2 \times 4 \times 10^5\,\text{g}}{\frac{1}{6} \times 10^{-23}\,\text{g}}.$$

Let us approximate $3 \times 2.2 \times 4 \times 6 = 150$. $131^2 \approx 1.7 \times 10^4$, and $150 \times 1.7 \approx 250$. So we are left with a reaction rate of

$$250 \times 10^{-4+7+4-45+5+23}\,\frac{\text{reactions}}{\text{sec}} = 2.5 \times 10^{-8}\,\frac{\text{reactions}}{\text{sec}}.$$

A year is about 3×10^7 seconds, so we predict only about one interaction per year. This is a hard game indeed. What this says is that this detector is sensitive to particles of such a small cross-section, but no smaller, at least for particles of this mass. If the dark matter were made of somewhat lower-mass WIMPs, we likely would have already discovered them with the Large Hadron Collider, the most powerful particle accelerator on Earth. Next generation particle detectors, using significantly more liquid xenon, will increase the chances of detecting the dark matter particles, if they are indeed WIMPs. The quest continues.

79. Rotating galaxies

79.a We are asking how much time it takes for an object moving across the sky at a speed of 200 km/sec at a distance of 30 million light-years to move an angle of 10 micro-arcseconds. Let's figure out what physical distance corresponds to 10 micro-arcseconds. A distance of 30 million light-years is roughly 10 million parsecs, so 1 arcsecond at that distance corresponds to 10 million AU, and 10 micro-arcseconds (10^{-5} arcsec) corresponds to 100 AU, or 1.5×10^{10} km. Traveling at 200 km/sec, it takes:

$$\frac{1.5 \times 10^{10} \, \text{km}}{200 \, \text{km/sec}} \approx 8 \times 10^7 \, \text{sec},$$

or about 3 years, to go that far. A measurement for a single star over the lifetime of Gaia (perhaps 6 years) would give a result right at the edge of detectability, but there are many stars in the galaxy that could be monitored, and combining the results of all of them could give an unambiguous detection of rotation.

We should mention that this problem is inspired by a now-famous (or infamous) measurement of the proper motions (i.e., apparent motion in the plane of the sky) of stars in the Pinwheel by the Dutch astronomer Adriaan van Maanen. Around 1915, he claimed that he had measured a nonzero proper motion for stars in the Pinwheel galaxy. Remembering that the proper motion (angle per unit time) is the physical speed in the plane of the sky divided by the distance, the fact that he was able to measure a proper motion implied that the distance to the Pinwheel wasn't very large, ruling out the idea that the Pinwheel was a distant "island universe," as some were claiming. This was used as a strong argument against the island universe hypothesis in the famous Great Debate between Curtis and Shapley in 1920. It later turned out that van Maanen's measurements were simply wrong, and that the island universe hypothesis was correct.

79.b This is a straightforward application of the small-angle formula, $\theta = s/d$. Here the angle is 10^{-5} arcseconds (which we will convert to radians below), $s = 2$ cm is the physical size of the object subtending the angle, and we are asked for the distance d:

$$d = \frac{s}{\theta} = \frac{2 \, \text{cm}}{10^{-5} \, \text{arcsec} \times \frac{1 \, \text{rad}}{2 \times 10^5 \, \text{arcsec}}} = 10^{10} \, \text{cm} = 400,000 \, \text{km}.$$

This distance is very close to the distance from Earth to the Moon. That is, 10 micro-arcseconds is the angle a dime on the Moon subtends, as seen from Earth.

80. Measuring the distance to a rotating galaxy

A 45° angle means that the motion in the plane of the sky (measurable with proper motion) and that along the line of sight (Doppler shift) are the same. That is, there is a little isosceles right triangle including all these motions, with sides equal to $1/\sqrt{2}$ of the hypotenuse. (Note that because spiral galaxies are inherently circular when seen face-on, their inclination can be measured directly by measuring their apparent projected shape on the sky. In particular, if it is tipped at $45°$, its shape will be elliptical on the sky, with a minor axis $1/\sqrt{2}$ as large as the major axis.)

Thus the physical speed of the Doppler shift (150 km/sec) and the proper motion are the same. The latter is 10 micro-arcseconds per year at distance d, and remembering that 1 arcsecond at 1 parsec corresponds to 1 AU, this is a speed of

$$v = 10^{-5}\,\frac{\text{arcsec}}{\text{year}} \times d \times \frac{1\,\text{year}}{3 \times 10^7\,\text{sec}} \times \frac{1\,\text{AU}}{1\,\text{arcsec} \times 1\,\text{pc}} \times \frac{1.5 \times 10^8\,\text{km}}{1\,\text{AU}} = 5 \times 10^{-5}\,\frac{d}{\text{pc}}\,\frac{\text{km}}{\text{sec}}.$$

This speed is equal to 150 km/sec, allowing us to solve for d (which, remember, was in units of parsec):

$$d = \frac{150}{5 \times 10^{-5}}\,\text{pc} = 30 \times 10^5\,\text{pc} = 3\,\text{Mpc}.$$

The Gaia satellite is hoping to make this type of measurement of nearby galaxies.

81. The Hubble constant

81.a The relationship between luminosity, distance, and brightness is given by the inverse square law:

$$\text{Brightness} = \frac{\text{Luminosity}}{4\pi\,\text{Distance}^2}.$$

Here we are given the luminosity of each galaxy (the four are the same, namely 4×10^{37} Joules/sec), and the brightness, in units of Joules/m^2/sec. Solving for the distance gives

$$\text{Distance} = \left(\frac{\text{Luminosity}}{4\pi\,\text{Brightness}}\right)^{1/2}.$$

Note that the numbers will come out in meters, which is just fine. It is straightforward in each case to divide by 3.1×10^{22} meters/Mpc, to get the distance in megaparsecs. When we do this for the four galaxies, we find:

$$\text{Galaxy 1: Distance} = 6.5 \times 10^{24}\,\text{meters} = 210\,\text{Mpc},$$

$$\text{Galaxy 2: Distance} = 8.4 \times 10^{24}\,\text{meters} = 270\,\text{Mpc},$$

$$\text{Galaxy 3: Distance} = 1.1 \times 10^{25}\,\text{meters} = 360\,\text{Mpc},$$

$$\text{Galaxy 4: Distance} = 1.5 \times 10^{25}\,\text{meters} = 490\,\text{Mpc}.$$

81.b The following is a table of the measured wavelengths for each of the two lines in each of the galaxies, the corresponding redshift from each of the lines, and the average redshift. The redshift is given by $z = (\lambda - \lambda_0)/\lambda_0$, where $\lambda_0 = 3935$Å and 3970Å for the two lines.

Galaxy	λ first line (Ångstrom)	λ second line (Ångstrom)	Redshift first line	Redshift second line	Average redshift
1	4,100	4,135	0.042	0.041	0.042
2	4,145	4,185	0.053	0.054	0.053
3	4,215	4,255	0.071	0.072	0.071
4	4,318	4,360	0.097	0.098	0.098

81.c The redshift is equal to the velocity of recession divided by the speed of light. So we can calculate the velocity of recession as the redshift times the speed of light. We'll work with velocities in units of kilometers per second, as we wish our final result for the Hubble constant to be in units of kilometers per second per megaparsec. The Hubble constant is given by the ratio of the velocity of each object to its distance. So we make another table:

Galaxy	Redshift	Velocity (km/sec)	Distance (Mpc)	$H_0 = V/D$ (km/sec/Mpc)
1	0.042	12,600	210	60
2	0.053	15,900	270	59
3	0.071	21,300	360	59
4	0.098	29,400	490	59

The four galaxies give consistent values of the Hubble Constant of about 60 km/sec/Mpc. Not identical to the modern value of 67 km/sec/Mpc, but close.

This calculation seemed quite straightforward; so why is there such controversy over the exact value of the Hubble constant? The difficult point is getting an independent measurement of the luminosity of each galaxy. The problem stated that each of the galaxies has the same luminosity as the Milky Way. That is only approximately true; the numbers were adjusted somewhat to make this come out with a reasonable value for H_0.

82. Which expands faster: The universe or the Atlantic Ocean?

Here we apply the expansion of the universe to the Atlantic Ocean. As the problem states, this is a bit of a mis-use of the Hubble Law, but it will give us a feel for how fast the expansion is. The speed at which a point 6,000 km away would recede, according to the Hubble Law, is

$$v = H_0 d.$$

To go on, we need to remember that the Hubble Constant needs to be put into useful units. We show in *Welcome to the Universe* that the reciprocal of the Hubble constant is the age of the universe, 14 billion years $\approx 5 \times 10^{17}$ sec. Thus H_0 is the reciprocal of this number, or 2×10^{-18}/sec. Plugging this value in, we find

$$v = H_0 d = 2 \times 10^{-18}\,\mathrm{sec}^{-1} \times 6,000\,\mathrm{km} \times 10^6\,\mathrm{mm/km} = 1.2 \times 10^{-8}\,\mathrm{mm/sec}.$$

Over the period of a year, the distance traveled is this times the number of seconds in a year, namely 3×10^7 sec, giving a distance of 4×10^{-1} mm, or 400 microns. Plate tectonics gives an enormously larger expansion, of 1 inch (about 25 millimeters) per year. That is, the Atlantic is spreading faster than the universe by a factor of $\frac{25\,\mathrm{mm}}{0.4\,\mathrm{mm}} \approx 60$. This means that the age of the Atlantic (since the time at which the Americas split off from Africa and Europe) is about 60 times less than the age of the universe. So, 14 billion years divided by 60 is about 200 million years, and the actual split between the continents is thought to have happened 170–200 million years ago. So our estimate is pretty good!

83. The third dimension in astronomy

Parallax refers to the apparent change in the position of stars as our perspective changes. In particular, we use it for the changing perspective as Earth goes around the Sun over the course of a year. The change in angle in the apparent position of a star is inversely proportional to the distance to the star; indeed, the distance in parsecs is the inverse of the parallax in arcseconds. The parallax for even the nearest stars is a bit less than 1 arcsecond. It is a challenge to measure parallax for stars more than a few hundred light-years away using telescopes on the ground, but the European Space Agency's Gaia satellite will be able to measure parallaxes for stars over a large portion of the Milky Way Galaxy.

The inverse square law relates brightness, luminosity, and distance of stars via the formula

$$b = \frac{L}{4\pi d^2}.$$

Thus if you can measure the brightness, and you know the luminosity, you can determine the distance. It is not always easy in astronomy to know the intrinsic luminosities of the objects you're observing! Standard candles are objects whose intrinsic luminosities you know or can infer from their observed properties. For stars, a measurement of the spectrum or color can allow an inference of the surface temperature, and knowing the relationship between temperature and luminosity on the main sequence then allows a determination of luminosity. However, it is not always obvious whether a given star is in fact on the main sequence. One can also use Cepheid variable stars, whose luminosities are directly related to the period over which they vary. This of course gets difficult for very distant (and thus faint) stars. Another type of standard candle used is Type Ia supernovae. Intervening dust will make a distant object appear fainter than it would otherwise be, making it difficult to apply the inverse square law.

Edwin Hubble discovered a direct proportionality between distance r and apparent recession velocities v of galaxies: $v = H_0 r$. The recession velocity is determined from the redshift measured from the spectrum. This works for galaxies; it is not valid for objects in our own Milky Way Galaxy. Of course, to apply this relation requires having an accurate value for the Hubble constant H_0, whose value has been controversial for quite a while. Astronomers are fairly confident now that the Hubble constant lies in the range 66–74 km/sec/Mpc, although controversy rages over the exact value in that range. Redshifts are measured as shifts from galaxy spectra; in particular, one measures the shift of specific spectral features, such as the pair of lines due to calcium absorption, from their rest wavelengths.

At very high redshifts, the simple form of the Hubble law breaks down, and one needs a specific cosmological model (which takes into account the expansion history of the universe) to relate redshift to distance. In fact, one needs to take care to define exactly what is meant by distance in an expanding universe; it is tricky, because the distance to a far-away galaxy continues to grow as the light from it makes its way to us.

84. Will the universe expand forever?

84.a Consider a sphere of radius r and density ρ. It has a volume $\frac{4}{3}\pi r^3$ and a mass $M = \frac{4}{3}\pi r^3 \rho$. The escape speed from a mass M of radius r is given by $\sqrt{\frac{2GM}{r}}$:

$$v_{\text{esc}} = \sqrt{\frac{8\pi G r^3 \rho}{3r}} = \sqrt{\frac{8\pi G r^2 \rho}{3}}.$$

84.b We need to compare the speed of the galaxy in question, $H_0 r$, with the escape speed calculated in part **a**. If the escape speed is larger than the actual speed, the galaxy will not escape and will eventually stop and fall back. That is, we want to ask about the conditions under which the following holds:

$$H_0 r < \sqrt{\frac{8\pi G r^2 \rho}{3}}.$$

Let us see what this condition this imposes on the density. We proceed by squaring both sides of the equation:

$$H_0^2 r^2 < \frac{8\pi G r^2 \rho}{3}.$$

There is a common factor of r^2 in this equation, so we can cancel it. Solving for density, we find

$$\rho > \frac{3 H_0^2}{8\pi G}.$$

This result is remarkable. Because the answer does not depend on either the mass or the distance of the galaxy in question, it must be true for all galaxies, and therefore is a statement about the universe as a whole.

That is, if the density of the universe is greater than the value given above (the so-called *critical density*), the gravity of the universe is great enough to stop the expansion and cause it to recollapse, ending in what is called the Big Crunch. If the density is equal to or less than the critical density, each galaxy is moving at its escape velocity or faster, and the universe will expand forever.

It is worth remarking that this calculation does not take into account the existence of dark energy. Dark energy is gravitationally repulsive and can prevent the universe from collapsing even if the density is larger than the critical value.

Our best measurements tell us that the actual mean density of the universe is about 30% of the critical value. So even without the existence of dark energy, the universe is fated to expand forever.

84.c First, let's calculate H_0:

$$H_0 = 67\,\mathrm{km\,sec^{-1}\,Mpc^{-1}} \times \frac{1{,}000\,\mathrm{m}}{1\,\mathrm{km}} \times \frac{1\,\mathrm{Mpc}}{3 \times 10^{22}\,\mathrm{m}} \approx 2 \times 10^{-18}\,\mathrm{sec^{-1}}.$$

(This is just the inverse of 14 billion years, expressed in seconds.)

So canceling the 3 and the π, the critical density is

$$\rho_{\mathrm{crit}} \approx \frac{H_0^2}{8G} \approx \frac{\left(2 \times 10^{-18}\,\mathrm{sec^{-1}}\right)^2}{8 \times \frac{2}{3} \times 10^{-10}\,\mathrm{m^3\,sec^{-2}\,kg^{-1}}} = \frac{4 \times 3}{16} \times 10^{-36+10}\,\mathrm{kg\,m^{-3}}$$

$$= 8 \times 10^{-27}\,\mathrm{kg\,m^{-3}}.$$

Remember that a hydrogen atom has a mass of 1.6×10^{-27} kg. Thus this density corresponds to about five hydrogen atoms every cubic meter. Even though this density appears tiny, if the universe were more dense than this, the self-gravity of the universe would be enough to counteract its expansion and would cause it to collapse some time in the future.

84.d We are given the mass of each galaxy and the distance between each galaxy. We can consider each galaxy as occupying a cube $r = 15$ million light-years on a side. Given the volume of this cube and the mass it encloses, we infer a density of

$$\frac{1\,\text{galaxy}}{(1.5 \times 10^7\,\mathrm{ly})^3} = \frac{10^{11}\,\mathrm{M_{Sun}} \times 2 \times 10^{30}\,\mathrm{kg/M_{Sun}}}{(1.5 \times 10^7\,\mathrm{ly} \times 10^{16}\,\mathrm{m/ly})^3} = \frac{2 \times 10^{41}\,\mathrm{kg}}{3 \times 10^{69}\,\mathrm{m^3}} \approx 7 \times 10^{-29}\,\mathrm{kg/m^3}.$$

So by this calculation, the density of the universe is less than 1% of the critical density; the universe will expand forever.

In fact, in this calculation, we have ignored the effects of the extended dark matter halos of galaxies, which causes us to significantly underestimate their masses. Our current best estimate for the density of the universe, including both ordinary matter and dark matter, is about 30% of the critical density. Dark energy comprises another 70%; the total mass-energy density of the universe is very close to the critical value. This is consistent with the predictions of the model stating that the universe underwent a period of inflation in the first tiny fraction of a second after the Big Bang.

85. The motion of the Local Group through space

85.a The Doppler formula tells us what we want to know:

$$\frac{v}{c} = \frac{\Delta I}{I} = 1.23 \times 10^{-3},$$

where v is the speed of Earth and I is the intensity of the CMB. Thus the speed is $1.23 \times 10^{-3} c = 360$ km/sec.

85.b First, let's add our answer in part **a** to 300 km/sec, to get a total of 660 km/sec, half of which is due to the Virgo Supercluster. The gravitational acceleration due to a mass M a distance R from us is GM/R^2. This acceleration, acting for a time t, induces a velocity

$$v = \frac{GMt}{R^2}.$$

In this case, we want to determine M, so we find

$$M = \frac{vR^2}{Gt}.$$

We have no clever scaling tricks to use here, so let's just calculate. Here v is half of 660 km/sec, or 330 km/sec, $R = 18$ Mpc, and $t = 14$ billion years $\approx 4 \times 10^{17}$ seconds is the age of the universe:

$$M = \frac{3.3 \times 10^7 \, \text{cm/sec} \times (18 \times 3 \times 10^{24} \, \text{cm})^2}{6.6 \times 10^{-8} \, \text{cm}^3/\text{g}/\text{sec}^2 \times 4 \times 10^{17} \, \text{sec}}.$$

Fortuitously, $3.3/6.6 = 1/2$; that's easy. Also, $18^2 \approx 300$, and $3^2 = 10$, so this becomes

$$M \approx \frac{10^7 \times 300 \times 10 \times 10^{48}}{8 \times 10^{-8+17}} \, \text{g} = \frac{3}{8} \times 10^{49} \, \text{g}.$$

Converting to solar masses gives us

$$M = 2 \times 10^{15} \, M_{\text{Sun}}.$$

85.c If this volume of space were at the average density, it would have a mass of half that we just calculated in part **b**: $10^{15} \, M_{\text{Sun}}$, or 2×10^{48} grams. This mass is contained in a sphere of radius 18 Mpc, or a volume of

$$V = \frac{4}{3}\pi(18 \times 3 \times 10^{24} \, \text{cm})^3.$$

$18 \times 3 = 54$, which when cubed is about 1.6×10^5. So the volume comes to $V = 6.4 \times 10^{77}\,\mathrm{cm}^3$, leading to a density of

$$\rho = \frac{M}{V} = \frac{2 \times 10^{48}\,\mathrm{g}}{6.4 \times 10^{77}\,\mathrm{cm}^3} \approx \frac{1}{3} \times 10^{-29}\,\frac{\mathrm{g}}{\mathrm{cm}^3}.$$

The critical density of the universe is about $10^{-26}\,\mathrm{kg/m}^3$, or $10^{-29}\,\mathrm{g/cm}^3$. So the density of the universe that we have calculated is close to, but somewhat less than, the critical density. The ratio of the two is

$$\Omega_{\mathrm{matter}} = \frac{\rho}{\rho_{\mathrm{crit}}} \approx \frac{1}{3}.$$

The calculation we've just done, refined to properly take into account the expansion of the universe, was one of the primary methods astronomers used in the 1980s to try to estimate the mass associated with the Virgo Supercluster, dark matter and all, and thus determine the overall density of the universe. We've gotten a value of $\Omega_{\mathrm{matter}} \approx 1/3$, in accord with the modern value of 0.30. When these calculations were first done, many were skeptical at the time: they were convinced that Ω_{matter} had to be 1 (to give a flat universe) and thus assumed that there was something wrong with the assumptions that went into the calculation. In fact, both sides were right: the conclusion that Ω_{matter} was significantly less than 1 was correct, and the discovery of dark energy in the late 1990s led to the conclusion that the universe is indeed flat.

86. Neutrinos in the early universe

86.a We need to start by calculating the number density of hydrogen atoms in the universe. We are told that it is 4% of the critical density, that is, 4% of 10^{-26} $\mathrm{kg/m}^3$, a density of 4×10^{-28} $\mathrm{kg/m}^3$. The mass of each hydrogen atom is 1.6×10^{-27} kg, so dividing this into the mass density gives us the number density for hydrogen. This gives us 0.25 atoms per cubic meter; one hydrogen atom in every 4 cubic meters.

We are told that neutrinos have a space density 1.5×10^9 times larger, or roughly 4×10^8 neutrinos per cubic meter (corresponding to 400 neutrinos per cubic centimeter). Neutrinos barely react with anything, so these neutrinos are streaming through your body as you read this!

86.b A little unit analysis tells us how to do this problem:

$$\text{Mass of neutrinos per unit volume} = \text{Number of neutrinos per unit volume}$$

$$\times \text{Mass per neutrino.}$$

Plugging in numbers, the mass density is

$$\text{Mass density} = \frac{4 \times 10^8\,\text{neutrinos}}{\mathrm{m}^3} \times \frac{0.1\,\text{electron volt}}{1\,\text{neutrino}} \times \frac{1.8 \times 10^{-36}\,\mathrm{kg}}{1\,\text{electron volt}}$$

$$= 7 \times 10^{-29}\,\frac{\mathrm{kg}}{\mathrm{m}^3}.$$

This is more than a factor of 100 below the critical density of the universe. Neutrinos have such low mass that despite their (relatively) large number density, they contribute negligibly to the density of the universe.

When people first realized that dark matter permeated the universe in the late 1970s and early 1980s, they did the early universe physics calculation that showed that neutrinos should be numerous, and speculated that dark matter was made of neutrinos. At the time, the constraint on the mass of neutrinos was weak enough to make this a plausible case. However, as the constraint on the neutrino mass has gotten lower, the calculation we've just done dashed people's hopes that the neutrinos were indeed massive enough to make up the dark matter.

87. No center to the universe

As far as we know, the universe has no edge. Given that the measured mass density of the universe is less than the critical value, our cosmological models tell us that the universe is infinite in extent, and therefore has no boundary, or edge. Without a boundary, one cannot define a center. Even if the mass density were above the critical density, the universe would be "closed"; analogous to the surface of a sphere, again there is no edge, and thus no center.

The spectra of the galaxies in the universe are redshifted, indicating that they are moving away from us. It therefore seems as if we are at the center of the expansion. However, as described in *Welcome to the Universe,* the Hubble law relating redshift and distance says that we could sit on *any* galaxy and come to the same conclusion. There is no unique center for the expansion. Alternatively, every point can equally well be considered the center of the expansion, as every point came from the infinitely dense Big Bang.

It is thus wrong to think of the Big Bang as exploding at a specific location and expanding into empty space. If the universe is infinite in extent now, it was at the time of the Big Bang as well, and it is the space itself that is expanding.

88. Luminous quasars

88.a We know that the Schwarzschild radius for 1 solar mass is 3 kilometers. We also know that the Schwarzschild radius is directly proportional to mass. So a black hole with a 10^8 times larger Schwarzschild radius must have 10^8 times larger mass, that is, 10^8 solar masses.

88.b The relation between luminosity, radius, and surface temperature for a blackbody is

$$L \propto R^2 T^4.$$

We want the temperature, so let's solve for that:

$$T \propto L^{1/4} \times R^{-1/2}.$$

We can write this relationship down for the blackbody in question and for the Sun, and take the ratio of the two:

$$\frac{T}{6,000\,\text{K}} = 10^3 \times \left(\frac{7 \times 10^5\,\text{km}}{3 \times 10^8\,\text{km}}\right)^{1/2} = 10^3 \times \left(\frac{1}{400}\right)^{1/2} = 1,000/20 = 50.$$

Thus, the temperature of the blackbody is 50×6000 K $= 3 \times 10^5$ K. In fact, the luminosity of the quasar doesn't come from a blackbody the size of the Schwarzschild radius, but rather from a hot accretion disk of gas. Friction in the disk causes the gas to spiral inward and fall into the black hole. The inner edge of the accretion disk has a radius of several times the black hole Schwarzschild radius.

88.c The inverse square law relates distance d, luminosity L, and brightness b:

$$b = \frac{L}{4\pi d^2}.$$

Here we are interested in distance, so we solve for it:

$$d \propto \left(\frac{L}{b}\right)^{1/2}.$$

We set up this equation for both the G star and the quasar, take the ratio, and recognizing that the brightnesses are the same, we have

$$\frac{d_{\text{quasar}}}{d_{\text{G star}}} = \left(\frac{L_{\text{quasar}}}{L_{\text{G star}}}\right)^{1/2} = (10^{12})^{1/2} = 10^6.$$

Thus

$$d_{\text{quasar}} = 10^6 \times 3{,}000 \text{ light-years} = 3 \times 10^9 \text{ light-years}.$$

This quasar is 3 billion light-years away, a truly cosmological distance.

89. The origin of the elements

The early universe was so hot that only protons (i.e., hydrogen nuclei), electrons, photons, neutrons, neutrinos (and presumably whatever elementary particle makes up the dark matter) could exist. In the first 3 minutes after the Big Bang, the neutrons combined with protons to make helium nuclei (roughly 25% by mass), along with trace amounts of deuterium and lithium nuclei. Indeed, the Big Bang predictions for how much helium and deuterium was created are in excellent agreement with observations and represent one of several strong experimental verifications of the hot Big Bang idea. However, no heavier elements were formed at this time.

Stars on the main sequence fuse hydrogen into helium in their cores. When the hydrogen runs out in the core, the star is no longer a main sequence star; it starts to collapse under gravity, and fuses helium to carbon and oxygen in its core, and the star becomes a red giant. For stars with mass less than eight solar masses, that's the full story, but for more massive stars, a continued series of nuclear reactions start up as each generation of fuel is exhausted in the core of the star, including synthesis of silicon, neon, and many other elements. The star develops an onion structure, with each layer burning a different element. Each stage lasts a shorter time, as the amount of energy that can be obtained from each element is less than the one before. When the core is pure iron, the star can no longer hold itself up. The reason is that iron is an element from which no further energy can be obtained from thermonuclear fusion; it is the most stable of the nuclei. The star collapses and explodes in a supernova; in the huge amount of energy released, atoms from the rest of the periodic table are generated. These elements are spewn out into the interstellar medium in the supernova explosion, to be incorporated into the next generation of stars and planets.

There are in fact additional mechanisms for making the elements beyond iron. Type Ia supernovae, in which a white dwarf star is pushed beyond the Chandrasekhar limit of 1.4 solar masses, are another means of making heavy elements. One way to make a white dwarf star so massive is by accretion of material from a companion star. Another way is to have a collision between a pair of white dwarfs, driven by the interaction with a third star in the system. Collisions between pairs of neutron stars are another mechanism for forging elements beyond iron.

In all these cases, the resulting explosion returns material that is otherwise in the core of the star to the interstellar medium, so that it may take part in the formation of future generations of stars and planets.

90. Lorentz factor

90.a When $v \ll c$, then $v/c \ll 1$, and y will be very close to unity. So let us write

$$y = \sqrt{1 - x^2} = 1 - \epsilon,$$

where $x = \frac{v}{c}$. Let us square both sides to get

$$y^2 = 1 - x^2 = (1 - \epsilon)^2.$$

Expanding out the right-hand side gives us

$$1 - x^2 = 1 - 2\epsilon + \epsilon^2.$$

As the problem hints, if ϵ is small, then ϵ^2 is tiny, and we can neglect it. This then gives

$$-x^2 = -2\epsilon,$$

or

$$\epsilon = \frac{1}{2} x^2.$$

Or, plugging in our expression for $x = \frac{v}{c}$:

$$\epsilon = \frac{1}{2} \left(\frac{v}{c}\right)^2.$$

Therefore, when $\frac{v}{c} \ll 1$, we can write

$$y = \sqrt{1 - \left(\frac{v}{c}\right)^2} \approx 1 - \frac{1}{2} \left(\frac{v}{c}\right)^2,$$

to a very good approximation.

Incidentally, this is an example of a general rule, which can be demonstrated with something called a *Taylor* series (which one learns about in calculus), namely:

$$(1 + x)^n \approx 1 + nx,$$

whenever the absolute value of x, $|x| \ll 1$.

90.b Note that $v^2/c^2 = (1 - \alpha)^2 = 1 - 2\alpha$, to a good approximation, ignoring the truly tiny α^2 term. Thus:

$$y = \sqrt{1 - \frac{v^2}{c^2}} \approx \sqrt{1 - (1 - 2\alpha)} = \sqrt{2\alpha} = \sqrt{2 \left(1 - \frac{v}{c}\right)}.$$

Here's another way to do the same calculation. We can recognize that the quantity under the square root is the difference of squares, and thus factorizes as

$$y = \sqrt{1 - \frac{v^2}{c^2}} = \sqrt{\left(1 - \frac{v}{c}\right)\left(1 + \frac{v}{c}\right)}.$$

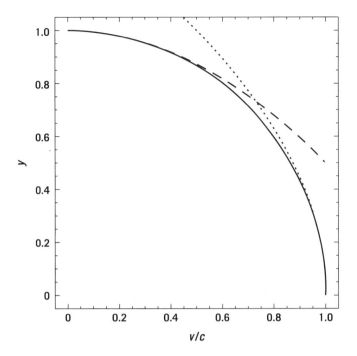

Figure 10. Figure for the solution to problem 90. The solid line is the exact Lorentz contraction formula as a function of the fraction of the speed of light, v/c. The dotted and dashed lines are the approximations worked out in parts **b** and **a**, for high v/c and low v/c, respectively.

Now, to a very good approximation, $1 + v/c = 2$ (after all, we're taking the limit of v approaching c), so this simply becomes $\sqrt{2(1 - v/c)}$, as above.

90.c The graph is shown in figure 10, giving both the exact formula for the Lorentz contraction and the two approximations we've just worked out. The approximations work impressively well. The overall shape of the curve is that of a piece of a circle, extending from unity at small speeds, to zero at the speed of light. It is within 10% of the approximation worked out in part **a** for speeds less than $0.76c$, and with the approximation worked out in part **b** for speeds above $0.65c$.

90.d Here we have to calculate the Lorentz factor for various values of v. For the very small values of velocity, we'll use the approximation of part **a**, and for the very large velocity, we'll use the approximation of part **b** (that's what the approximations are for!). For $v = 0.6c$ and $0.8c$, we'll do the full calculation, and also show the results for the two approximations; we'll see that both do a pretty good job!

- $v = 100$ km/hour. Let's first convert to meters per second; with 1,000 meters in a kilometer and 3,600 seconds in an hour, this is roughly 30 meters per second, or $10^{-7}c$, to one significant figure. By the approximation we worked out in part **a**, then,

$$y \approx 1 - \frac{1}{2}\frac{v^2}{c^2} = 1 - 5 \times 10^{-15} = 0.999999999999995.$$

Again, this is the factor by which time slows down for a driver driving down the highway; it is so close to unity as to be completely unnoticeable. Note also that most calculators can't do the exact calculation, which is one reason that the approximation we derived in part **a** is important.

It is worth thinking about the significant figures here. You might argue that because the velocity is given to a single significant figure, you should give y to the same significance (i.e., 1). But what makes y interesting is the way in which it differs from 1 (that's the bit that is distinctive about special relativity); in this sense, the quantity that should be listed to a single significant figure is $1 - y$.

- $v = 30$ km/sec, or $10^{-4}c$: $y = 1 - \frac{1}{2}\left(\frac{v}{c}\right)^2 = 1 - \frac{1}{2} \times 10^{-8} = 1 - 0.000000005 = 0.999999995$.
- $v = 0.6c$: $y = \sqrt{1 - 0.6^2} = \sqrt{1 - 0.36} = \sqrt{0.64} = 0.8$. For the small-$v$ approximation, we get $y = 1 - \frac{1}{2}(0.6^2) = 1 - \frac{1}{2} \times 0.36 = 0.82$; pretty close! For the large-v approximation, we get $y = \sqrt{2(1 - 0.6)} = \sqrt{0.8} \approx 0.9$; not as close to the right answer.
- $v = 0.8c$: $y = \sqrt{1 - 0.8^2} = \sqrt{1 - 0.64} = \sqrt{0.36} = 0.6$. For the small-$v$ approximation, we get $y = 1 - \frac{1}{2}(0.8^2) = 1 - \frac{1}{2} \times 0.64 = 0.68$, a bit high. For the large-v approximation, we get $y = \sqrt{2(1 - 0.8)} = \sqrt{0.4} \approx 0.63$; pretty good!
- $v = 0.9999995c$: $y = \sqrt{2(1 - 0.9999995)} = \sqrt{2 \times 5 \times 10^{-7}} = \sqrt{10^{-6}} = 10^{-3}$.

It it worth remarking on the last of these values. Time can slow down by a factor of 1,000 if you are traveling sufficiently close to the speed of light. The next problem will explore a physical situation in which such enormous speeds are relevant.

91. Speedy muons

Look at problem 90.**d**, where we calculated the Lorentz contraction factor for this value of the speed; it is 10^{-3}. That is, time for the muon progresses 1,000 times slower than it does for us. Thus as seen by us, the muon now has a lifetime 1,000 times *longer*, that is, 2.2×10^{-3} seconds (i.e., the lifetime gets divided by the Lorentz contraction factor). How far does it go at almost the speed of light? Well, light travels at 1 foot per nanosecond, and its lifetime is 2.2 million nanoseconds, so it goes 2.2 million feet, or about 700 kilometers.

If it weren't for the special relativistic effects, it would have its usual lifetime of 2.2 microseconds, and would therefore be able to travel 1/1,000 as far (i.e., about 700 meters).

Thus the observation of muons that have traveled 100 kilometers is a direct demonstration of special relativity at work; there would be virtually no cosmic ray muons from the top of Earth's atmosphere at Earth's surface if Einstein were not correct.

92. Energetic cosmic rays

Given the expression for the total energy of the particle, the value of y in this case is the ratio of the rest-mass energy of the proton to the total energy:

$$\frac{mc^2}{mc^2/y} = y.$$

Plugging in numbers yields

$$y = \frac{10^9\,\text{eV}}{3 \times 10^{20}\,\text{eV}} \approx 3 \times 10^{-12}.$$

Wow, that is astonishingly small! By our calculation in problem 90.**b**, the quantity y is equal to $\sqrt{2(1 - v/c)}$, so we can solve for v:

$$\sqrt{2\left(1 - \frac{v}{c}\right)} = 3 \times 10^{-12}$$

$$2\left(1 - \frac{v}{c}\right) = 10^{-23}$$

$$\frac{v}{c} = 1 - 5 \times 10^{-24}$$

$$v = c\left(1 - 5 \times 10^{-24}\right).$$

Here's another way to do the problem, which just moves the place where we do the approximation. We write

$$\sqrt{1 - \left(\frac{v}{c}\right)^2} = y.$$

Solving for v gives

$$1 - \left(\frac{v}{c}\right)^2 = y^2,$$

$$v = c\sqrt{1 - y^2}.$$

The value of y is tiny, so y^2 is minuscule. We can use the same sort of approximation we derived in problem 90.**a** to write this as

$$v = c\left(1 - \frac{1}{2}y^2\right).$$

(Make sure you followed the algebra in this last step!) When we plug in numbers, we find $y^2 \approx 10^{-23}$, and so the velocity is

$$v = c\left(1 - 5 \times 10^{-24}\right),$$

just as we found above.

So this particle is moving at almost, but not quite, the speed of light. At these high speeds, the interesting quantity, which one can actually measure, is the energy of the particle, which gives us an estimate of y. Given how close the speed of such a particle is to the speed of light, we can't measure the difference between its speed and the speed of light by seeing how long it takes to travel a certain distance.

Einstein has taught us that no object can go faster than the speed of light in a vacuum, and indeed, no massive object (i.e., anything with a rest mass) can move *at* the speed of light, because at the speed of light, $y = 0$, and the energy of the particle, mc^2/y, would have to be infinite. So how is it that photons can go at the speed of light? They have no rest mass; that is, if you stop a photon, it ceases to exist, and gives up its energy. A photon can travel only at the speed of light.

So how do you detect a cosmic ray like the Oh-My-God particle of such amazingly high energy? When such a cosmic ray hits the top of the atmosphere, it collides with atoms, and its kinetic energy is converted (via $E = mc^2$) into enormous numbers of new particles of all sorts. Those new particles have plenty of kinetic energy as well, and they collide with atoms to create further particles. Pretty soon, the initial cosmic ray creates a cascading shower of particles, which will hit the ground over a large area (many square kilometers). Astronomers have built a number of arrays of detectors, spread over wide areas to catch the particles that reach the ground from these cosmic ray showers.

93. The Titanic is moving

93.a The contraction of lengths due to special relativity for an object of length l traveling at speed v is $l\sqrt{1 - v^2/c^2}$, where c is the speed of light. We are asked what v must be for the ship to shrink from 882.5 feet to 882.4 feet. We could simply plug this into the calculator and solve for v, but let's see if we can do this without a calculator:

$$882.4 = 882.5\sqrt{1 - \frac{v^2}{c^2}}.$$

We know that because the contraction is quite small, the speed will be appreciably smaller than the speed of light, that is, $v^2/c^2 \ll 1$. Under these conditions,

$$\sqrt{1 - \frac{v^2}{c^2}} = 1 - \frac{1}{2}\frac{v^2}{c^2},$$

to a very good approximation. You can convince yourself of this by taking the square of both sides and realizing that if v^2/c^2 is small, v^4/c^4 is minuscule! (See problem 90 for the details.) Using this approximation, we can write

$$882.4 = 882.5\left(1 - \frac{v^2}{2c^2}\right).$$

Subtracting 882.4 from both sides and rearranging gives

$$\frac{0.1}{882.5} = \frac{v^2}{2c^2},$$

$$2 \times 10^{-4} = \frac{v^2}{c^2}.$$

Taking the square root of both sides gives us

$$v = 1.4 \times 10^{-2}c = 1.4 \times 10^{-2} \times 3 \times 10^5 \,\text{km/sec} \approx 4{,}500 \,\text{km/sec}.$$

Thus the *Titanic* would have to be traveling at 4500 kilometers per second for us to notice a 1-inch contraction in its length! This is why we don't notice the effect; to get even this tiny contraction would require speeds orders of magnitude higher than those we ever experience in everyday life.

93.b To calculate the energy of the explosion resulting from the impact of the *Titanic* as it hits the ground after being dropped from a height of 51.3 km, we first assume that the kinetic energy of the *Titanic* when it hits the ground will be entirely transformed into 'explosion' energy. The energy of the explosion is therefore

$$E_{\text{explosion}} = E_{\text{kinetic}} = \frac{1}{2}mv^2 = 0.5 \times \left(45{,}000\,\text{tons} \times 10^3\,\frac{\text{kg}}{\text{ton}}\right)\left(1\,\text{km/sec} \times \frac{10^3\,\text{m}}{1\,\text{km}}\right)^2$$

$$= 2.25 \times 10^{13} \,\text{Joules}.$$

We know that 1 kiloton of TNT, when it explodes, produces an energy of 4.2×10^{12} Joules. The energy of the Titanic explosion in equivalent kilotons of TNT is therefore

$$E_{\text{explosion}} = \frac{2.25 \times 10^{13}\,\text{Joules}}{4.2 \times 10^{12}\,\text{Joules/kiloton}} \approx 5 \text{ kilotons of TNT}.$$

Boom!

93.c Using Einstein's equation $E = mc^2$, the amount of mass that would have to be annihilated to create a similarly energetic explosion is

$$m = \frac{E_{\text{explosion}}}{c^2} = \frac{2.25 \times 10^{13} \, \text{Joules}}{(3 \times 10^8 \, \text{m/sec})^2} = 2.5 \times 10^{-4} \, \text{kg} = 0.25 \, \text{grams}.$$

That is about the mass of a bug.

94. Aging astronaut

As seen by us, the time will be

$$\frac{\text{Distance}}{\text{Speed}} = \frac{4 \, \text{ly}}{0.8 \, c} = 5 \, \text{years}.$$

But her clock will tick slower, by a factor $\sqrt{1 - v^2/c^2} = 0.6$. Thus she will have aged 3 years.

Another way solve this problem: from her perspective, the 4 light-year distance is contracted by the Lorentz factor of 0.6; thus she concludes that the two stars flying by her at 0.8 times the speed of light are separated by a distance of only $0.6 \times 4 = 2.4$ ly. That separation distance passing by her at 0.8 times the speed of light takes

$$\frac{2.4 \, \text{ly}}{0.8c} = 3 \, \text{years}.$$

95. Reunions

Your roommate has been traveling at close to the speed of light, so much so that his clock has been ticking slower by a factor of 40 years/50 years = 4/5. This factor is equal to that of the Lorentz factor term in special relativity, $\sqrt{1 - v^2/c^2}$. Let us solve for v:

$$\sqrt{1 - \frac{v^2}{c^2}} = 4/5$$

$$1 - \frac{v^2}{c^2} = 16/25$$

$$\frac{v^2}{c^2} = 9/25$$

$$\frac{v}{c} = 3/5.$$

Thus he was traveling at 60% of the speed of light, or 180,000 km/sec.

In our reference frame, he was traveling for 50 years at 60% of the speed of light, and therefore traveled 30 light-years. He made the round trip; therefore, the star is 15 light-years away.

96. Traveling to another star

96.a From our perspective, to travel 20 light-years at 80% of the speed of light takes

$$\frac{20 \, \text{ly}}{4/5 \times 1 \, \text{ly/year}} = 25 \, \text{years}.$$

96.b From the point of view of the Gliesians, their time is contracted by the Lorentz factor:

$$\sqrt{1 - \frac{v^2}{c^2}} = \sqrt{1 - 0.8^2} = \sqrt{0.36} = 0.6.$$

Thus the time they think it takes is

$$25 \, \text{years} \times 0.6 = 15 \, \text{years}.$$

96.c The Gliesians will see the distance between their planet and Earth as shrunk by the same factor of 0.6, and thus will report it as

$$20 \, \text{ly} \times 0.6 = 12 \, \text{ly}.$$

Alternatively, they will see the separation between their planet and Earth whizzing past them at $0.8c$, giving a distance

$$15 \, \text{years} \times 0.8 \times 1 \, \text{ly/year} = 12 \text{ light-years.}$$

97. Clocks on Earth are slow

The Lorentz formula involves multiplying a time span by the factor $\sqrt{1 - v^2/c^2}$. Here, the speed in question is 30 km/sec, which is *much* less than the speed of light. We saw in problem 90 that, in the limit of small speeds, the Lorentz factor is well approximated by $1 - \frac{1}{2}\frac{v^2}{c^2}$. That is, for a time t, the difference between the two clocks is

$$t - t\sqrt{1 - \frac{v^2}{c^2}} \approx t\left[1 - \left(1 - \frac{1}{2}\frac{v^2}{c^2}\right)\right] = \frac{t}{2}\left(\frac{v}{c}\right)^2.$$

We want to know: for what value of the time t does this difference equal 1 second? $v = 30$ km/sec is 10^{-4} of the speed of light, so we get

$$\frac{t}{2} \times (10^{-4})^2 = 1 \, \text{sec,}$$

$$t = 2 \times 10^8 \, \text{sec} \times \frac{1 \, \text{year}}{3 \times 10^7 \, \text{sec}} \approx 6 \, \text{years.}$$

98. Antimatter!

First, antimatter has mass just like ordinary matter; it doesn't have "negative mass" in any sense of the term. So it feels gravity and is equivalent to energy in the usual $E = mc^2$ sense. What makes it *anti*matter is just the sense described in the problem, namely when it comes into contact with ordinary matter, it annihilates completely. Nasty stuff to have around....

Let's start this problem algebraically, and plug in numbers in the end. We have two trucks, each of mass m, each traveling at speed v. The kinetic energy of each is $\frac{1}{2}mv^2$, so the two of them together have a total kinetic energy of mv^2.

Now, the matter and antimatter trucks each have a rest-mass energy of mc^2 (same m as above!), and thus the two together have a total energy of $2mc^2$, all of which will be released upon collision. Wow! The problem asks: what is the ratio of these two energies? That's straightforward:

$$\text{Ratio} = \frac{2mc^2}{mv^2} = 2\left(\frac{c}{v}\right)^2.$$

Note that the mass drops out of this problem. OK, let's plug in numbers:

$$\text{Ratio} = 2 \left(\frac{3 \times 10^5 \, \text{km/sec}}{100 \, \text{km/hour} \times 1 \, \text{hour}/3600 \, \text{sec}} \right)^2 = 2 \times (3,000 \times 3,600)^2$$

$$\approx 2 \times 10^{14}.$$

(In the last step, we approximated $3 \times 3.6 \approx 10$.) That is, the collision of the matter and antimatter trucks is 200 trillion times more energetic than the head-on collision of two ordinary trucks, each traveling at highway speeds. You definitely had better run if you see such a collision about to happen! Luckily, there is not much antimatter around (if there were some, it would instantly annihilate with whatever ordinary matter is nearby, thereby destroying it). It is actually a real cosmic mystery why the universe did not create equal amounts of matter and antimatter in the Big Bang. It is a good thing, too: if equal amounts had been created, it would all have annihilated, leaving a universe of lots of energy (in the form of photons) and no material objects. A rather boring place!

Note that we didn't need to know the masses of the trucks to calculate the ratio. But for completeness, let's just work out how much energy this matter-antimatter collision actually releases: $2mc^2 = 2 \times 10^4 \, \text{kg} \times (3 \times 10^8 \, \text{m/sec})^2 = 2 \times 10^{21}$ Joules. Note that our Sun produces 200,000 times this much energy every second. That is, every second, the Sun turns roughly 4 million tons of matter into pure energy (via the process of nuclear fusion, not matter-antimatter annihilation).

99. Energy in a glass of water

Water's chemical composition is H_2O, with an atomic weight of 2 from the hydrogen and 16 from the oxygen. Thus the fraction of the mass of water that is in hydrogen is $2/18 \approx 1/10$. That is, an 8-ounce glass of water contains about $1/40$ of a kilogram of hydrogen, or 0.025 kilograms.

We also know that 0.7% of the mass of hydrogen is converted to energy when thermonuclear fusion takes place. Thus, the mass of that glass of water that is converted to energy is roughly $0.025 \text{kg} \times 0.007 \approx 2 \times 10^{-4}$ kg. This is converted to energy via $E - mc^2$:

$$E = 2 \times 10^{-4} \, \text{kg} \times (3 \times 10^8 \, \text{m/sec})^2 \approx 2 \times 10^{13} \text{ Joules}.$$

That's a seriously big number! Let's compare with the energy usage of New York City. Multiplying the number of residents of the city by their energy usage per person gives a total energy usage of 5×10^{10} kilowatt-hours per year. A kilowatt is 1,000 watts, or 1,000 Joules per second, and a kilowatt-hour is therefore

$$1 \, \text{kilowatt-hour} = 1,000 \, \frac{\text{Joules}}{\text{sec}} \times \text{hour} \times \frac{3,600 \, \text{sec}}{\text{hour}} = 3.6 \times 10^6 \, \text{Joules}.$$

Therefore, the total energy usage of New York, per year, is

$$5 \times 10^{10} \, \frac{\text{kilowatt-hour}}{\text{year}} \times 3.6 \times 10^6 \, \frac{\text{Joules}}{\text{kilowatt-hour}} \approx 2 \times 10^{17} \, \frac{\text{Joules}}{\text{year}}.$$

However, that's the rate of energy usage per year. We are asked for the rate per day, so we divide the above by 365 days to get a daily energy use rate of 5×10^{14} Joules by all the residents of Manhattan. With a single glass of water giving you 2×10^{13} Joules, it will only take about 25 glasses of water per day to keep all of Manhattan powered. At this level, we're not going to run out of water any time soon! So if we can figure out how to create conditions similar to those at the core of the Sun here on Earth so as to coax hydrogen into fusing, we would have an inexhaustible source of energy.

100. Motion through spacetime

Earth moves in a helix through spacetime. The space part is circular (well, really slightly elliptical), but it moves forward in the time dimension, stretching the circle into a helix. Indeed, because the speed of light is much greater than the speed of Earth around the Sun (30 kilometers per second, or 10^{-4} the speed of light), this is a very stretched out helix; Earth is moving forward much faster in time than in space. See figure 18.1 of *Welcome to the Universe* for an illustration.

101. Can you go faster than the speed of light?

Consider a spaceship traveling faster than the speed of light; we will show a contradiction with the postulates of special relativity. The lightbeam travels at the speed of light by the second postulate. The front wall of the spacecraft is going faster, and has a head start, so the light beam will never hit that front wall. You will infer from this fact that something is wrong, and conclude that you must be traveling faster than the speed of light, in violation of the first postulate, which states that there is no experiment that you can perform that will show you are moving.

102. Short questions in special relativity

102.a TRUE. Simultaneity is a relative concept in special relativity, and depends on one's reference frame.

102.b The equation is $E = mc^2$. The three terms are E, energy; m, mass; and c, the speed of light. The equation implies that energy and mass are equivalent, and can be transformed into each other. This transformation process goes on in atomic bombs and in thermonuclear reactions in the centers of stars, for example.

102.c Of course, we are all traveling to the future, at a rate of 1 second every second! Here we ask the circumstances under which some can travel faster into the future than others. If we travel close to the speed of light, our clock will tick slower than that of someone who stands still. We will therefore travel to the future faster than the person standing still. We know this experimentally from observations of atmospheric muons, and from putting accurate atomic clocks on high-speed airplanes. The greatest time traveler to date is Russian cosmonaut Gennady Padalka, who has traveled forward in time by 1/44 of a second relative to the rest of us on Earth.

102.d Yes, they both could be telling the truth. Observers moving at different speeds can disagree on the order of events, and in particular, whether two events are simultaneous.

102.e The answer is no. A proper spacetime diagram will show the worldlines of the astronaut, the front of the ship, and the back of the ship from the perspective of someone standing still, in addition to the worldlines of the two photons. See figure 18.2 of *Welcome to the Universe.*

103. Tin Can Land

103.a Figure 11 shows the solution to this problem. It shows three geodesics. There are other geodesics connecting the square and the triangle in addition to the ones shown here; they wrap around the can multiple times.

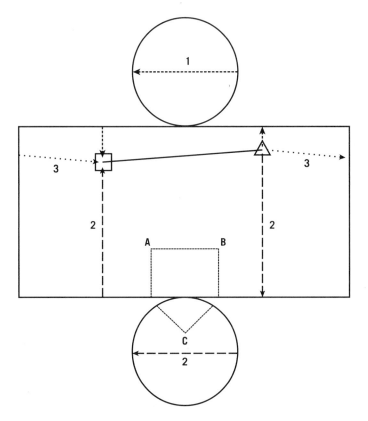

Figure 11. Solution to various parts of problem 103.

103.b The shortest geodesic is the one going over the top (labeled 1 in figure 11).

103.c The triangle ABC is shown in the figure; there are many triangles one can draw with three right angles, all of which will cross the boundary at the top or bottom of the can.

103.d There is (infinite) curvature at the boundaries between the side of the can and its top; this geometry is not Euclidean! Although the curvature is confined to an infinitesimal region, if you cross it, you can have interesting deviations from Euclidean predictions, such as a triangle with three right angles. In general, the rules of Euclidean geometry hold only in flat space. In this case, the space is locally flat over most of the space, but crossing those regions in which it is not flat causes these deviations from the Euclidean rules. The angles of a triangle drawn fully on the side, or on either the top of the bottom of a tin can, will add up to 180°, because the triangle lies completely within a region that is locally flat.

104. Negative mass

Both Einstein and Newton would agree that the ball would hit the floor. Newton would say: let the mass of the ball be m. It feels a gravitational force from Earth (of mass M and radius r) of GMm/r^2. Equating this to its mass times its acceleration gives us $a = GM/r^2$, independent of the mass m, even if m is negative. This acceleration is downward, so the ball would fall downward.

Einstein would say: objects in a gravitational field follow geodesics, independent of their mass. To put it another way, a uniform gravitational field is equivalent to an accelerated reference frame; the floor comes up to meet this ball, independent of its mass.

It is worth going back to the Newtonian language. Note that mass enters in two different places: there is the *inertial* mass associated with the relationship between force and acceleration ($F = ma$), and the *gravitational* mass in Newton's law of gravitation. Einstein's Equivalence Principle works, because those two masses are the same and thus cancel each other out in the expression for acceleration. It is not obvious a priori that these two forms of mass are the same, and scientists have developed very clever experiments (google "Eötvös experiment") that have confirmed that inertial and gravitational masses are in fact the same to the limits of our current ability to measure them.

Negative-mass balls as described in this problem do not actually exist in the universe as far as we know; nevertheless, both Newtonian gravity and general relativity give us a language to talk about them. Antimatter, however, is a real physical phenomenon and appears in particle physics collisions of various sorts. It is antimatter in the sense that if it comes into contact with ordinary matter, the two annihilate, turning all their joint mc^2 into energy. Antimatter has positive mass. People have done Eötvös experiments on antimatter particles, and have demonstrated that the Equivalence Principle holds for them as well: antimatter falls at the same rate in a gravitational field as does ordinary matter.

105. Aging in orbit

105.a We are given the rate at which a clock at radius r ticks relative to one infinitely far away. Here we are asked to compare the rate of a clock at radius r relative to one at the radius of Earth (i.e., on Earth's surface). We can think about this by considering the rate of each of these clocks relative to a distant clock. That is, the clock on Earth's surface has a rate a factor

$$\sqrt{1 - \frac{1}{c^2} \frac{2GM_\oplus}{r_\oplus}}$$

slower than the distant clock, while the clock at radius r has a rate that is slower by a factor

$$\sqrt{1 - \frac{1}{c^2} \frac{2GM_\oplus}{r}}.$$

Note that both these expressions are less than 1, but because $r > r_\oplus$, the stationary clock at radius r ticks faster than that at Earth's surface. Indeed, the relative rate of the two is just the ratio of these two expressions:

$$\sqrt{\frac{1 - \frac{1}{c^2} \frac{2GM_\oplus}{r}}{1 - \frac{1}{c^2} \frac{2GM_\oplus}{r_\oplus}}}.$$

Again, this expression is greater than 1.

105.b The velocity of an astronaut in a circular orbit is something you have seen in previous problems. Circular motion at speed v at a radius r gives rise to an acceleration v^2/r, which we know is due to gravity. Thus if the astronaut has a mass m, Newton's second law tells us:

$$F = ma$$

$$\frac{GM_\oplus m}{r^2} = m\frac{v^2}{r}.$$

Solving for v gives

$$v = \sqrt{\frac{GM_\oplus}{r}}.$$

The time-slowing factor due to special relativity is the now-familiar $\sqrt{1 - v^2/c^2}$, which in this case gives

$$\sqrt{1 - \frac{GM_\oplus}{rc^2}}.$$

Note again how similar this looks to the expression above for time dilation due to gravity. Again, this is the rate that an orbiting clock at radius r ticks relative to a stationary clock at the same radius.

105.c In part **a**, we calculated the ratio of rates of stationary clocks at radius r and $r_\oplus$ (due to general relativity), while in part **b**, we calculated the ratio of the rates of an orbiting clock at radius r to a stationary clock at the same radius. Thus the ratio of the rate of an orbiting clock at radius r to a stationary one on the ground is simply the product of these two results:

$$\sqrt{\frac{1 - \frac{1}{c^2}\frac{2GM_\oplus}{r}}{1 - \frac{1}{c^2}\frac{2GM_\oplus}{r_\oplus}} \times \left(1 - \frac{1}{c^2}\frac{GM_\oplus}{r}\right)}.$$

This is a complex expression, and there is no obvious way to simplify it without making some approximations, as we will do in the next part.

105.d Let's follow the hint. If we square both sides of the equation:

$$\sqrt{1-x} \approx 1 - \frac{1}{2}x,$$

we find

$$1 - x \approx 1 - 2 \times \frac{1}{2}x + \frac{1}{4}x^2.$$

If x is small, $\frac{1}{4}x^2$ is really small, and we can ignore it, so the right-hand side becomes $1-x$, which clearly equals the left-hand side. If the square of both sides of our expression are equal, then the original expression must be true. Note that there was a similar calculation in problem 90.

105.e Let's do the first of these approximations. Following the hint, let us write

$$\frac{1}{1-x} \approx 1 + x,$$

and do algebraic manipulations on this to get to a result we know is true. If we are successful in doing so, we can be confident that the original result was correct. Multiplying both sides by $1-x$ gives

$$1 \approx (1+x)(1-x) = 1 - x^2.$$

But we know $x \ll 1$, so x^2 is really small and can be neglected relative to 1. So the above equation gives us the clearly true statement, $1 \approx 1$. That establishes the first relation.

The second relation follows similarly:

$$(1 - x)(1 - y) = 1 - (x + y) + xy.$$

But if both x and y are much less than 1, their product is again really small and can be neglected. Thus,

$$(1 - x)(1 - y) \approx 1 - (x + y),$$

as we were to show.

105.f Now we will simplify the expression we derived in part **c**, using the various tricks we've assembled in parts **d** and **e**. We will show at the end that the approximation that the various relevant pieces are much smaller than 1 is correct. We will do this in pieces, starting with the computation from part **a**. We can write

$$\sqrt{\frac{1 - \frac{1}{c^2}\frac{2GM_\oplus}{r}}{1 - \frac{1}{c^2}\frac{2GM_\oplus}{r_\oplus}}} \approx \sqrt{\left(1 - \frac{1}{c^2}\frac{2GM_\oplus}{r}\right) \times \left(1 + \frac{1}{c^2}\frac{2GM_\oplus}{r_\oplus}\right)},$$

using the first expression derived in part **e**. We then use the second expression derived in part **e** to get

$$\sqrt{1 - \left[\frac{GM_\oplus}{c^2}\left(\frac{2}{r} - \frac{2}{r_\oplus}\right)\right]}.$$

This then gets multiplied by the expression in part **b**; again we use the second expression of part **d** to simplify it to

$$\sqrt{1 - \left[\frac{GM_\oplus}{c^2}\left(\frac{3}{r} - \frac{2}{r_\oplus}\right)\right]}.$$

We're almost done! Using the tools developed in part **d**, we know how to simplify expressions that look like $\sqrt{1 - x}$, where x is small. Our expression becomes

$$1 - \left[\frac{GM_\oplus}{2c^2}\left(\frac{3}{r} - \frac{2}{r_\oplus}\right)\right].$$

That is our final answer.

However, we do need to justify the use of the various approximations we've made. We dealt with a variety of expressions of the form $1 - x$; in every case, x is of the form: $\frac{GM_\oplus}{rc^2}$. The smallest r we considered (and therefore the largest the expression $\frac{GM_\oplus}{rc^2}$ is) is $r = r_\oplus$. So let's plug in numbers at $r = r_\oplus$, and see what we get:

$$\frac{GM_\oplus}{r_\oplus c^2} = \frac{\frac{2}{3} \times 10^{-10}\,\mathrm{m^3\,sec^{-2}\,kg^{-1}} \times 6 \times 10^{24}\,\mathrm{kg}}{6.4 \times 10^6\,\mathrm{m} \times (3 \times 10^8\,\mathrm{m/sec})^2},$$

where we have been careful to put all numbers into MKS units. To simplify this equation, let us approximate the radius of Earth as 6×10^6 meters, and we'll write $3^2 \approx 10$. A lot of stuff cancels, and we get a final value of 7×10^{-10}. Note that this quantity has no units. This is indeed a number much much smaller than 1!

105.g Here we ask for the radius at which the expression we derived in part **f** is equal to unity. This clearly holds when the quantity in brackets is equal to zero:

$$\frac{3}{r} - \frac{2}{r_\oplus} = 0,$$

or

$$r = 1.5 r_\oplus.$$

Given the radius of Earth, 6,400 km, this is at a distance of 9,600 km from Earth's center, or 3,200 km above Earth's surface. The expression in part **f** is less than 1 at smaller radii, so astronauts on the Space Shuttle age less (albeit by a tiny amount) than those staying home.

We have referenced all these calculations relative to people on the surface of Earth. Of course, the Earth (and the spacecraft orbiting it) are orbiting around the Sun at 30 kilometers per second. It would be a separate calculation to determine the time-slowing effects of that orbit, for example, relative to somebody sitting still, far from the Sun.

106. Short questions in general relativity

106.a TRUE. This is the basis of general relativity, which dispenses with the idea of gravity as a force in the Newtonian sense. Rather, mass-energy causes spacetime to curve, and bodies follow geodesics (analogous to straight lines) in that curved spacetime.

106.b A geodesic is the path that a particle subject to no forces (other than gravity) takes through spacetime. Equivalently, it is the straightest path between two points in curved spacetime.

106.c Einstein's insight was that a uniform gravitational field is indistinguishable from an accelerated reference frame, and therefore the two could be considered to be equivalent. He termed this the Equivalence Principle (what he called the happiest idea of his life), and it led to his general theory of relativity.

106.d The first was the precession of the perihelion of Mercury. General relativity predicted the precession at exactly the amplitude observed. The second was the deflection of light from the Sun. In 1919, this was observed during a full solar eclipse and was again found to be in good agreement with the general relativity predictions. More recent tests include the observation of gravitational lenses, whereby the light of a distant galaxy or quasar is split into multiple images by the gravitational effect of a foreground galaxy, and the dramatic detection of gravitational waves from the merging of a pair of black holes.

107. A black hole in the center of the Milky Way

107.a Newton's form of Kepler's third law tells us that the mass in solar masses is

$$M(\text{solar masses}) = \frac{a^3(\text{AU})}{P^2(\text{years})}.$$

Plugging in numbers yields

$$M = \frac{10^9}{250} = 4 \times 10^6 \, M_{\text{Sun}}.$$

107.b Remember that 1 AU subtends 1 arcsecond at a distance of 1 parsec. Thus the angle in question, in arcsec, is 1,000 AU/8,000 parsec = 1/8 arcsec, or $0.125''$. A more difficult way to do this would be to convert AU and parsecs (or light-years) to kilometers, take the ratio, and convert the resulting angle from radians to arcsec. This is easier to do if you remember that 1 pc = 200,000 AU, and there are 200,000 arcsec in a radian.

107.c The radius of a black hole is proportional to its mass. A 1 solar-mass black hole has a radius of 3 kilometers, so a $4 \times 10^6 \, M_{\mathrm{Sun}}$ black hole has a radius of $3 \times 4 \times 10^6 \, \mathrm{km} = 1.2 \times 10^7$ km.

108. Quick questions about black holes

108.a The answer is yes, via Hawking radiation, whereby a virtual pair of particles at the event horizon is split, one falling into the black hole and the other being emitted. Hawking radiation is very weak, however, and hasn't yet been experimentally verified. The very massive black holes that are found at the centers of galaxies, with masses of more than a billion solar masses, will take more than 2×10^{94} years to decay via Hawking radiation.

108.b He coined the term "black hole." He got Bekenstein thinking about entropy in black holes, which led to the concept of a black hole giving off radiation, and with Feynman, he explored the idea that a positron is an electron going back in time (equivalently, that all positrons and electrons in the universe are the same particle).

108.c TRUE. Massive black holes (the kind made in the collapse of stars, or the supermassive black holes that reside at the centers of galaxies) give off Hawking radiation very slowly, so slowly that Hawking radiation has never been experimentally verified. Hawking radiation carries away energy, and thus mass, and thus an isolated black hole will have a finite lifetime.

108.d FALSE. Black holes emit Hawking radiation, and thus decay (albeit very slowly). While Hawking radiation is a straightforward prediction of our understanding of quantum mechanics and general relativity, it has never been observed. A pair of merging black holes will also release energy (and thus mass) in the form of gravitational radiation. This was first observed in 2015, when 3 solar masses were lost to gravitational radiation in the merger of two black holes.

109. Big black holes

Note that the Schwarzschild radius for a black hole is directly proportional to its mass, so if we work out the numbers for a standard mass black hole, say, 1 solar mass, it will be easy to scale the result for any other mass. For 1 solar mass:

$$R_{\mathrm{Schwarzschild}} = \frac{2GM_{\mathrm{bh}}}{c^2},$$

$$R_{\mathrm{Schwarzschild}} = \frac{2 \times \frac{2}{3} \times 10^{-10} \, \mathrm{m^3 \, sec^{-2} \, kg^{-1}} \times 2 \times 10^{30} \, \mathrm{kg}}{(3 \times 10^8 \, \mathrm{m \, sec^{-1}})^2} \approx \frac{1}{3} \times 10^4 \, \mathrm{m}$$

or about 3 kilometers. Again, the Schwarzschild radius is proportional to the mass, so we can simply multiply this result by the number of solar masses for each black hole.

And for the tidal acceleration, at $d = 2R_{\mathrm{Schwarzschild}}$,

$$\mathrm{Tidal\,acceleration} = \frac{2GM_{\mathrm{bh}}r}{d^3} = \frac{2GM_{\mathrm{bh}}r}{(2 \times \frac{2GM_{\mathrm{bh}}}{c^2})^3} = \frac{c^6 r}{32G^2 M_{\mathrm{bh}}^2} = \frac{c^6 r}{32G^2 M_{\mathrm{Sun}}^2} \left(\frac{M_{\mathrm{Sun}}}{M_{\mathrm{bh}}} \right)^2.$$

In the last step, we've scaled to the case of the mass of the Sun. If we plug in numbers, we find

$$\text{Tidal acceleration} = \frac{(3 \times 10^8 \, \text{m sec}^{-1})^6 \times 1 \, \text{m}}{32 \times (\frac{2}{3} \times 10^{-10} \text{m}^3 \, \text{sec}^{-2} \, \text{kg}^{-1})^2 \times (2 \times 10^{30} \, \text{kg})^2} \left(\frac{M_{\text{Sun}}}{M_{\text{bh}}} \right)^2.$$

$$\text{Tidal acceleration} = 1.3 \times 10^9 \left(\frac{M_{\text{bh}}}{M_{\text{Sun}}} \right)^{-2} \text{m sec}^{-2}.$$

All that is left now is to substitute the masses (in solar masses) of the different objects we've been asked to consider. We find:

Mass	Radius	Tidal acceleration
30 M_{Sun}	90 km	1.4×10^6 m/sec^2
$4 \times 10^6 M_{\text{Sun}}$	1.2×10^7 km	8×10^{-5} m/sec^2
$10^9 M_{\text{Sun}}$	3×10^9 km	1.3×10^{-9} m/sec^2

It is very interesting to notice that black holes with masses somewhat larger than the Sun have huge tidal forces that could break an object within a few Schwarzschild radii to pieces in fractions of a second. Compare the tidal acceleration we've calculated with the acceleration of gravity here on Earth (10 m/sec^2); you're being pulled apart by more than 100,000 g's! At the other extreme, a $10^9 M_{\text{Sun}}$ black hole at the center of a quasar has a Schwarzschild radius of 3 billion meters, or 20 AU, but tiny tidal forces. In fact, if we were to fall into a Schwarzschild black hole of mass $10^9 M_{\text{Sun}}$, we would certainly survive well after crossing the Schwarzschild radius (we still would be unable to escape, however!). But soon after crossing the event horizon at the Schwarzschild radius, the tidal forces would get larger and larger, and would tear us apart before we disappeared into the singularity in the middle.

110. A Hitchhiker's challenge

110.a The Schwarzschild radius of a black hole of mass M is

$$R_{\text{Sch}} = \frac{2GM}{c^2}.$$

The volume of a sphere of this radius is just the familiar $\frac{4}{3}\pi R_{\text{Sch}}^3$. The density is the mass divided by the volume, giving

$$\text{Density} = \frac{M}{\frac{4}{3}\pi \left(\frac{2GM}{c^2} \right)^3} = \frac{3c^6}{32\pi G^3 M^2}.$$

A messy expression! The more massive the black hole is, the smaller the density becomes. Thus there is a mass at which the black hole has the density of paper, which is what we are trying to figure out. If you gather together enough paper to equal that mass, it will collapse to form a black hole.

It is worth pointing out that the actual material that makes up the singularity at the center of the black hole is much smaller than its Schwarzschild radius and at much higher density. The material is, as best we understand, crushed to a point, of extent of order the Planck length.

110.b The density is the mass per unit volume. If we can figure out the volume of a square meter of paper (whose mass we know, 75 grams), we can calculate its density.

The volume of a piece of paper is its area times its thickness. The thickness is 0.1 millimeter, or 10^{-4} meters, and so the volume of a square meter of paper is $1\,\text{meter}^2 \times 10^{-4}\,\text{meter} = 10^{-4}\,\text{meter}^3$. Therefore, the density of paper is

$$\rho = \frac{7.5 \times 10^{-2}\,\text{kg}}{10^{-4}\,\text{m}^3} = 7.5 \times 10^2\,\text{kg/m}^3,$$

similar to (but slightly less than) the density of water (remember, paper is made of wood, and wood floats in water!).

110.c Here we equate the expression for density we found in part **a** with the density we calculated in part **b**, and solve for mass. Let's first do it algebraically:

$$\rho = \frac{3\,c^6}{32\,\pi G^3 M^2},$$

$$M = \sqrt{\frac{3\,c^6}{32\,\pi G^3 \rho}}.$$

Now let's plug in numbers. This will be fun without a calculator:

$$M = \sqrt{\frac{3\,(3 \times 10^8\,\text{m/sec})^6}{32\,\pi (2/3 \times 10^{-10}\text{m}^3\,\text{sec}^{-2}\,\text{kg}^{-1})^3 \times 7.5 \times 10^2\,\text{kg/m}^3}} \approx 3 \times 10^{38}\,\text{kg},$$

where we made all the usual approximations of $\pi = 3$, $3^2 = 10$, and so on. We need to express this in solar masses, so we divide by 2×10^{30} kg (i.e., 1 solar mass) to get $1.5 \times 10^8\,M_{\text{Sun}}$ as our final answer. A black hole 150 million times the mass of the Sun has the same density as a piece of paper!

Surely this is science fiction! Well, yes, *The Hitchhiker's Guide to the Galaxy* is indeed science fiction, but do such incredibly massive black holes actually exist? Indeed they do: the cores of massive galaxies (including our own Milky Way) do contain enormous black holes. One of the most massive such black holes known to exist is in the core of a particularly luminous galaxy known as Messier 87, with an estimated mass of 3 billion solar masses.

We still have to calculate the Schwarzschild radius of the black hole. We could plug into the formula for a Schwarzschild radius and calculate away, but we prefer a simpler approach. We know the Schwarzschild radius is proportional to the mass of a black hole, and we know that a 1 solar-mass black hole has a Schwarzschild radius of 3 kilometers. So a 150-million-solar-mass black hole has a Schwarzschild radius 150 million times larger, or 4.5×10^8 kilometers. We are asked to express this in terms of AU; 1 AU is 1.5×10^8 kilometers, so the Schwarzschild radius of such a black hole is 3 AU.

110.d We know the entire mass of the black hole. If we can calculate the mass of a single piece of paper, the ratio of the two gives the total number of pages. So let's calculate the mass of a single piece of paper. We know that a square meter of paper has a mass of 75 grams. How many square meters is a standard-size sheet? One inch is 2.5 centimeters $= 2.5 \times 10^{-2}$ meters. So $8.5 \times 11\,\text{inch}^2 \approx 100\,\text{inch}^2 \approx 6 \times 10^{-2}\,\text{meter}^2$. Thus the mass is:

$$\text{Mass of a piece of paper} = 7.5 \times 10^{-2}\,\text{kg/m}^2 \times 6 \times 10^{-2}\,\text{m}^2 \approx 5 \times 10^{-3}\,\text{kg}.$$

That is, a piece of paper weighs about 5 grams. We divide this into the mass we calculated above:

$$\text{Number of sheets of paper} = \frac{\text{Mass of rule book}}{\text{Mass per page}} = \frac{3 \times 10^{38}\,\text{kg}}{5 \times 10^{-3}\,\text{kg/page}} = 6 \times 10^{40}\ \text{pages.}$$

(Strictly speaking, if the rule book is printed on both sides of the page, we should multiply the above result by a factor of 2.) That is one seriously long set of rules!

Finally, note that because the density of a more massive black hole is smaller, as we saw in part **a**, the above mass and number of pages of the Brockian Ultra Cricket rule book is really just a lower limit. That is, if the rule book were even larger than what we've just calculated, it would still collapse into a black hole.

Thanks to Alexandro Strauss (Michael's son, who has *The Hitchhiker's Guide to the Galaxy* practically memorized) for finding the quote about Brockian Ultra Cricket.

111. Colliding black holes!

111.a The orbit of each black hole is a circle of radius equal to its Schwarzschild radius $R_{\text{Schwarzschild}}$, and thus the circumference of the orbit is equal to 2π times that. Its period is that distance, divided by the speed of light:

$$\text{Period} = 0.004\,\text{sec} = \frac{2\pi \times R_{\text{Schwarzschild}}}{c}.$$

The speed of light is 300,000 kilometers per second, and using our usual approximation of $\pi = 3$, this simplifies to:

$$R_{\text{Schwarzschild}} = 200\ \text{kilometers.}$$

We know that the Schwarzschild radius corresponding to a solar mass is about 3 kilometers, and we have a number about 60 times larger. So each black hole has a mass of 60 solar masses. In fact, a more accurate calculation using the equations of general relativity shows that we're off by a factor of 2: the two black holes each have a mass of about 30 solar masses.

111.b The luminosity is the energy (3 solar masses times c^2), divided by the time (0.02 second):

$$L = \frac{3 \times 2 \times 10^{30}\,\text{kg} \times (3 \times 10^8\,\text{m/sec})^2}{0.02\,\text{sec}} \approx 3 \times 10^{49}\ \text{watts.}$$

Note that because we used MKS units consistently, the answer must be in watts.

111.c This is a straightforward calculation: 10^{11} solar luminosities per galaxy, and 10^{11} galaxy total, gives a total luminosity of 10^{22} solar luminosities. Multiplying by the luminosity of the Sun, about 4×10^{26} watts, gives a total luminosity of 4×10^{48} watts, a factor of 10 less than the value we calculated in part **b**. That is, for 1/50 of a second, these colliding black holes put out about 10 times more luminosity than all the stars, in all the galaxies, of the entire observable universe. That is awesome!

111.d The strain is the ratio of the change in distance divided by the distance (4 kilometers) itself, so the change of distance is the strain times 4 kilometers. Remembering that there are 10^{10} Ångstroms in a meter, and 1,000 meters in a kilometer, this gives us a change of distance of $4 \times 10^{-21+10+3}$ Ångstrom $= 4 \times 10^{-8}$ Ångstrom. This is roughly 4×10^{-3} times the radius of a proton. It is astonishing that we can measure such a tiny signal! It is even more amazing that we can interpret it as a pair of colliding black holes, which very briefly outshone the entire rest of the observable universe!

112. Extracting energy from a pair of black holes

112.a Each nonrotating black hole has a spherical event horizon with a Schwarzschild radius of $r_{\text{Sch}} = \frac{2GM}{c^2}$ and thus a surface area

$$A = 4\pi r_{\text{Sch}}^2 = 4\pi \left(\frac{2GM}{c^2}\right)^2.$$

There are initially two such holes so the total initial area of their event horizons is twice that:

$$A_{\text{total,initial}} = 2A = 32\pi \left(\frac{GM}{c^2}\right)^2.$$

112.b The area of the final black hole's event horizon must be greater than the total initial value $A_{\text{total,initial}}$ we found in part **a**:

$$A_{\text{final}} = 4\pi r_{\text{Sch,final}}^2 > 32\pi \left(\frac{GM}{c^2}\right)^2.$$

But the Schwarzschild radius of the final black hole is related to its mass using the usual formula:

$$4\pi \left(\frac{2GM_{\text{final}}}{c^2}\right)^2 > 32\pi \left(\frac{GM}{c^2}\right)^2.$$

Canceling common factors on both sides of the equation gives us:

$$M_{\text{final}}^2 > 2M^2,$$

or

$$M_{\text{final}} > \sqrt{2}M.$$

112.c The initial mass was $2M$, and the final mass must be larger than $\sqrt{2}M$, thus the amount of mass lost must be less than the difference between the two, that is, $(2-\sqrt{2})M$. If the mass is lost in the form of gravitational radiation, the upper limit on the amount of energy carried by gravitational waves is that times c^2, or $(2-\sqrt{2})Mc^2$.

112.d The two black holes are initially very far apart, so their initial gravitational potential energy is negligible. Thus, the total mass-energy of the system initially is just $E = 2Mc^2$, and the fraction of that released in the form of gravitational waves is smaller than

$$\frac{(2-\sqrt{2})Mc^2}{2Mc^2} = \frac{(2-1.41421\ldots)}{2} = 0.29289.$$

Thus, less than 29.3% of the initial mass-energy of the two black holes is turned into gravitational waves during their collision. In practice, the amount is usually considerably less than this, depending on the details of the collision. As we learned in problem 111, the LIGO experiment observed gravitational waves emitted by the collision of a 29-solar-mass black hole and a 36-solar-mass black hole to produce a

62 solar mass black hole. Thus, "only" 3 solar masses, or about 5% of the initial mass, was converted to energy in the form of gravitational waves.

113. Quick questions about time travel

113.a Given our current understanding, a time machine cannot take you back in time before the moment at which the time machine itself was built. No time machine has been built yet, so no time tourists can visit the present or the historical past.

113.b You need two cosmic strings passing each other at close to the speed of light; by making a loop around the two of them when they come close to each other, you can come back before you started. The space around each of the cosmic strings is conical, or pizza shaped, which means that you can find a path around them that beats a light beam.

114. Time travel tennis

The solution is shown in figure 12. Note that the ball is a jinn particle; its worldline is closed like a circle and does not have endpoints.

115. Science fiction

There is of course no correct answer here, but here is a grading rubric for this assignment.

If the science is more-or-less correct and the story is compelling, give it an A.

If the story just doesn't work, the science has obvious flaws, or is really not core to the story, give it a B.

If there seems absolutely no effort put into the story at all, give it a C.

If the story is brilliant, and you want to show it to all your friends, give it an A+.

116. Mapping the universe

The correct order is

- The Hubble Space Telescope
- Earth's Moon
- The WMAP satellite
- Jupiter
- Kuiper Belt Objects
- The star Alpha Centauri
- The star Sirius
- The center of the Milky Way
- M31, the Andromeda galaxy
- The Sloan Great Wall of galaxies
- The highest redshift galaxy known
- The Cosmic Microwave Background

117. Gnomonic projections

117.a The center of a great circle is the center of the sphere; indeed, that is one of the defining characteristics of a great circle. Thus, consider the plane in which the great circle lies: it includes the center of the circle. The gnomonic projection of the great circle is the shadow of the great circle onto a plane, cast by a light at the center of the sphere. This then is the intersection of the plane, including both the great circle and the center of the sphere, with the plane on which the projection lies.

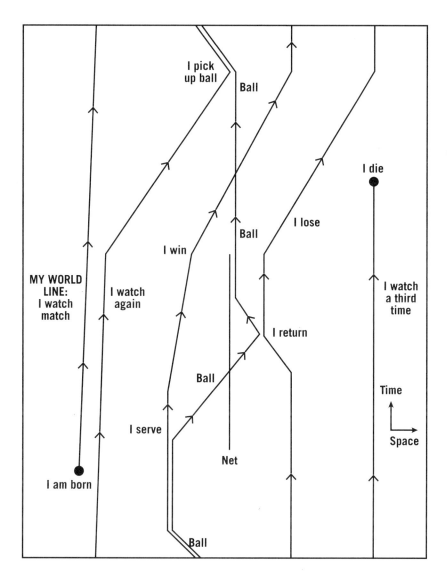

Figure 12. Solution for problem 114.

The intersection of two planes is a straight line, as you learned in your high school geometry class.

It is not correct to say that a geodesic ("straightest line between two points") in one projection should be identical to a geodesic in any other. Consider, for example, the line between New York and Tokyo; the great circle projects onto a curved line in a Mercator projection (the standard rectangular projection that is used for maps of the world in grade school classrooms).

Incidentally, the unusual term, "gnomonic" refers to a "gnomon," an old word for the part of a sundial that casts a shadow, as the gnomonic projection is not dissimilar to the projection of the shadow of the tip of the gnomon onto the planar face of a sundial. The word "gnomon" derives from the Greek root for "knowing."

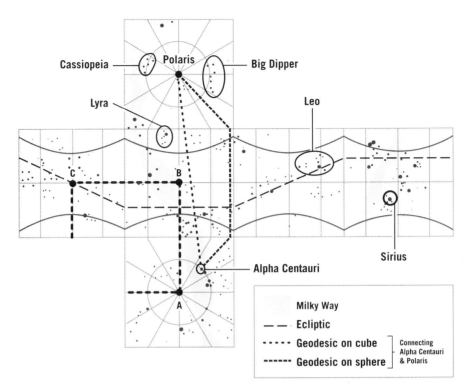

Figure 13. Demonstrating answers to various parts of problem 117.

117.b Consider a plane in front of the sphere. There is a hemisphere closest to the plane; it is clear that the light from this hemisphere will be projected onto the plane. The light going through the opposite hemisphere is projected in the opposite direction and will never make it to the plane in question. The great circle marking the boundary between the two hemispheres is, formally speaking, projected to the plane at an infinite distance (this is why we make the reference to slightly less than one hemisphere).

117.c The answer to these and subsequent questions are indicated in figure 13.

117.d The ecliptic is shown in figure 13. The ecliptic is a great circle; thus it projects to a straight line in each of the panels of the projection. But at the interstices between the panels, it bends sharply.

117.e The problem here is that both the ecliptic and the Celestial Equator are great circles, and therefore intersect (at two points). Yet they map onto straight lines in the gnomonic projection, which appear in some of the squares to be parallel. What's going on? The solution is simple: the straight lines do not remain straight as we go from one panel to another. Each panel is a separate gnomonic projection, and as we've already seen when we drew the ecliptic, the straight lines that a great circle projects onto need not be straight across the boundaries from one panel to another. So the laws of Euclidean geometry only hold on one panel at a time. Note that the Celestial Equator is a special case; because it is a line of symmetry relative to how we placed the planes for the gnomonic projections, it is indeed a straight line all the way through.

Note that it is not obvious a priori that the ecliptic and the Celestial Equator should be parallel at all in a gnomonic projection. It so turns out that they are in two of the panels, which is a consequence of the specific arrangements of the panels. The deeper paradox (solved again, by the statements above) is that two great circles on a sphere always cross each other twice, while two straight lines in Euclidean space can cross only zero or one time.

117.f Shown in figure 13. It is the line of longitude (a great circle) from Alpha Centauri to the North Pole. Lines of longitude fan out from the South Pole on the south circumpolar chart. Start from Alpha Centauri and draw a line straight away from the South Pole (at the center of the chart), until it hits the edge of the south circumpolar chart. The line of longitude then becomes a vertical straight line on the summer chart. It then enters the north circumpolar chart and goes straight to the North Celestial Pole at the center of the chart.

117.g No, it is not the same as the great circle in part **f**. The reason is that the geometry on the sphere and on the projection cube are different. See figure 13. It is a straight line on the folded-out star charts.

117.h The particular angles shown happen to be right angles in this gnomonic projection (although in general, the gnomonic projection does not preserve angles). So the angles indeed add up to 270 degrees. However, this is not a contiguous triangle in the gnomonic projection onto the cube. As we saw in part **e**, funny things happen when you cross the boundary between the different planes. In this case, the triangle has a break in it when one opens the cube to lay the six panels flat as in figure 13, which is why its angles don't follow the usual Euclidean rules.

118. Doctor Who in Flatland

118.a There are two sorts of vertices to examine here: those associated with the top corners of the sink, and those associated with the bottom. At the base of the sink, we have three 90° faces coming together at a vertex, for a total of $3\pi/2$, and thus a deficit of $\pi/2$, a positive quantity. At the top corner of the sink, we have $270°(=3\pi/2)$ on the flat surface, and $2 \times 90° = \pi$ associated with each of the two sink side faces, adding up to a total angle of $5\pi/2$. We are told that the mass is 2π radians minus this total angle of $5\pi/2$ radians, namely, $-\pi/2$. Thus these vertices each require a negative mass.

118.b There are four positive masses, each of mass $+\pi/2$, and four negative masses, each of mass $-\pi/2$. Adding them up gives us zero mass!

118.c The total mass associated with the Tardis is 0, and thus the spacetime outside the Tardis is not curved, but resembles a flat plane. There is no net mass in the center, so the exterior looks like a flat infinite plane, as we see.

119. Quick questions about the shape of the universe

119.a TRUE. However, Kaluza-Klein theory turned out to be equivalent to general relativity and electromagnetism as originally formulated, and thus made no new predictions. No experiment could be done to distinguish it from general relativity and electromagnetism.

119.b The prediction in question was the existence of the cosmic microwave background; it is important because it was a successful prediction of the Big Bang model. Indeed, the Big Bang model makes a whole series of predictions for the nature of the cosmic microwave background, all of which are in beautiful accord with observations:

- Its temperature
- The fact that follows the blackbody spectral form extremely accurately
- The fact that it is isotropic
- The fact that the power spectrum of fluctuations (minute deviations from isotropy) follow a specific form, as a function of scale, predictable in detail from the tenets of Big Bang theory and inflation.

119.c They are as follows:

- An open universe: the density is less than the critical density. It will expand forever, and has a geometry with negative curvature.
- A flat universe: it has density equal to the critical density. It will expand forever but will asymptotically approach zero expansion velocity. It has a flat geometry.
- A closed universe: it has density greater than the critical density. It will recollapse sometime in the future. It has a positive curvature geometry.

119.d The sum is greater than 180°. This is a curved geometry whose space part is a three-dimensional analog to the two-dimensional surface of a sphere. Like the surface of a sphere, the angles of a triangle add up to more than 180°. In an open Friedmann universe (whose two-dimensional analog is a saddle shape), the angles of a triangle add up to less than 180°. In the absence of a cosmological constant, an open Friedmann model has a mass density less than the critical density, while the closed model has a mass density greater than the critical density. If the density is equal to the critical value, the geometry is Euclidean (i.e., the angles of a triangle add up to 180°, as you learned in high school geometry).

119.e The introduction of the cosmological constant to his field equations of general relativity, according to a conversation related by George Gamow. He introduced the term because the equations otherwise predicted that the universe should be either expanding or contracting. Had he believed his original equations, he could have predicted that this was the case. More than a decade later, Edwin Hubble discovered that universe was in fact expanding.

120. The earliest possible time

120.a The wavelength λ of a photon is related to the frequency ν by $\lambda \nu = c$. But we also know that $h\nu = E = Mc^2$. Thus the Compton wavelength is:

$$\lambda_{\text{Compton}} = \frac{c}{\nu} = \frac{ch}{E} = \frac{ch}{Mc^2} = \frac{h}{Mc}.$$

120.b We equate the expression for the Compton wavelength and the Schwarzschild radius:

$$\frac{h}{Mc} = \frac{2GM}{c^2}.$$

Cross-multiplying gives:

$$hc^2 = 2GM^2c \rightarrow M = \sqrt{\frac{hc}{2G}}.$$

What we have just derived is called the "Planck mass." Max Planck derived this first, after he had first come up with his eponymous constant around 1900, from "dimensional analysis." General relativity and black holes were not yet even a figment in Einstein's imagination, and Compton's work was also more than a decade into the future, so Planck took a different approach. He knew that his constant (which he first developed in his attempt to understand blackbody radiation) was a fundamental physical constant, along with the speed of light and Newton's gravitational constant. He realized that he could combine them to end up with a quantity with units of mass, and wondered what its physical significance could be.

(We should mention that there are several different conventions for defining the Planck mass; the most common incorporates another factor of π under the square root, giving $M_{\mathrm{Planck}} = \sqrt{\frac{hc}{2\pi G}}$.)

OK, let's plug in numbers, working in cgs units:

$$M_{\mathrm{Planck}} = \sqrt{\frac{\frac{2}{3} \times 10^{-26}\,\mathrm{g\,cm^2/sec} \times 3 \times 10^{10}\,\mathrm{cm/sec}}{2 \times \frac{2}{3} \times 10^{-7}\,\mathrm{cm^3/sec^2/g}}}.$$

The $\frac{2}{3}$ cancels (it is just a coincidence that h and G both have a decimal part close to $\frac{2}{3}$), leaving

$$M_{\mathrm{Planck}} = \sqrt{1.5 \times 10^{-26+10+7}\,\mathrm{g^2}} = \sqrt{15 \times 10^{-10}}\,\mathrm{g} \approx 4 \times 10^{-5}\,\mathrm{g},$$

or about 40 micrograms. That may not sound very impressive, but in particle physics, it is the highest mass an elementary particle could have. Any higher, and it would collapse to form a black hole.

To convert this to GeV, we remember that a proton has a rest-mass energy of about a GeV, so the number of GeV in a Planck mass is roughly equal to the number of proton masses in a Planck mass:

$$\frac{M_{\mathrm{Planck}}}{M_{\mathrm{proton}}} \approx \frac{4 \times 10^{-5}\,\mathrm{g}}{\frac{1}{6} \times 10^{-23}\,\mathrm{g/GeV}} \approx 2 \times 10^{19}\,\mathrm{GeV}.$$

This is a lot of energy indeed; the energy to which the Large Hadron Collider accelerates its particles is only a few thousand GeV. Even the "Oh-My-God" particle of problem 92 had an energy of a "mere" 3×10^{11} GeV.

120.c The Planck length is derived from the expression for the Schwarzschild radius:

$$\mathrm{Planck\ length} = \frac{2\,G M_{\mathrm{Planck}}}{c^2} = \sqrt{\frac{4\,G^2}{c^4}\frac{hc}{2G}} = \sqrt{\frac{2\,hG}{c^3}}.$$

The Planck time is the time for light to travel that distance (i.e., this quantity divided by c):

$$\mathrm{Planck\ time} = \frac{\mathrm{Planck\ length}}{c} = \sqrt{\frac{2\,hG}{c^5}}.$$

(Like the Planck mass, the standard definition of Planck length and Planck time differ slightly from these expressions; Planck length $= \sqrt{\frac{hG}{2\pi c^3}}$ and Planck time $= \sqrt{\frac{hG}{2\pi c^5}}$.) We could calculate directly from these equations, but let's take a different approach by scaling from what we know about the Schwarzschild radius for 1 solar mass. The Schwarzschild radius is proportional to mass, and is 3 km $= 3 \times 10^5$ cm for the Sun, so for a Planck mass of 4×10^{-5} g, the Schwarzschild radius is

$$\text{Planck length} = 3 \times 10^5 \, \text{cm} \times \frac{4 \times 10^{-5} \, \text{g}}{2 \times 10^{33} \, \text{g}} = 6 \times 10^{5-5-33} \, \text{cm} = 6 \times 10^{-33} \, \text{cm}.$$

This is an incredibly small distance. Here's one way to think about it. Consider the size of a hydrogen atom: about 1 Ångstrom, or 10^{-8} cm. The Planck length is more than 24 orders of magnitude smaller still. Thus the Planck length is to an atom as an atom is to

$$10^{-8} \, \text{cm} \times 1.6 \times 10^{24} = 1.6 \times 10^{16} \, \text{cm} \approx 1,000 \, \text{AU}.$$

So with our current understanding of the laws of physics, we cannot speak about spatial scales any smaller than the Planck length. People speculate that if we could somehow magnify space to resolve such scales, we would find it to be "foamy" with extreme curvature (take a look at the article about "quantum foam" in Wikipedia).

The Planck time is the time for light to travel that distance, namely, the Planck length divided by the speed of light, or 2×10^{-43} seconds. This represents the earliest time that we can push back our understanding of the Big Bang. It is also the shortest time interval we can measure on any clock, as described at the beginning of chapter 24 of *Welcome to the Universe*.

121. The worst approximation in all of physics

121.a Let's first do this algebraically. We found expressions in problem 120 for the Planck mass and Planck length. We can create a density out of this in the obvious way:

$$\rho_{\text{Planck}} = \frac{M_{\text{Planck}}}{L_{\text{Planck}}^3}.$$

(We're doing an order-of magnitude calculation, and so are not worrying here about factors of $4\pi/3$.) But L_{Planck} is given by the Schwarzschild formula:

$$L_{\text{Planck}} = \frac{2GM_{\text{Planck}}}{c^2},$$

so

$$\rho_{\text{Planck}} = \frac{M_{\text{Planck}} c^6}{8 \, G^3 M_{\text{Planck}}^3} = \frac{c^6}{8 \, G^3 M_{\text{Planck}}^2} = \frac{c^6}{8 \, G^3} \frac{2 \, G}{hc} = \frac{c^5}{4 \, G^2 h}.$$

We could plug in numbers at this stage, but instead, but we'll calculate it from our knowledge of the numerical values of the Planck mass and length from problem 120:

$$\rho_{\text{Planck}} = \frac{4 \times 10^{-5} \, \text{g}}{(6 \times 10^{-33} \, \text{cm})^3} \approx \frac{4}{200 \times 10^{-94}} \frac{\text{g}}{\text{cm}^3} = 2 \times 10^{92} \frac{\text{g}}{\text{cm}^3}.$$

That's an impressively high density! Far higher than any other density we have seen in astronomy.

121.b The critical density of the universe is about 10^{-26} kg/m^3, or 10^{-29} g/cm^3. Thus the dark energy density is 70% of that critical density, or about 7×10^{-30} g/cm^3. Let's take the ratio of this density to that we calculated in part **a**:

$$\frac{2 \times 10^{92}}{7 \times 10^{-30}} \approx 3 \times 10^{121}.$$

Wow. This has famously been called "the worst approximation in all of physics." That is, one assumes that the dark energy is due somehow to physics related to the interface of quantum mechanics and general relativity; based on this, one determines roughly what the density of this dark energy should be. The answer you get is off by 121 orders of magnitude! This was a spectacularly unsuccessful calculation!

Clearly there is more to the story than this. One of the many ways to express our puzzlement about dark energy is why its density is as small as it is. Given that something causes its density to be so small relative to its "natural" value (i.e., the Planck density we've just calculated), the next question is, why is its value not exactly zero? This is just another way of saying that we really don't understand what dark energy is all about.

If the dark energy really were at the Planck value, the universe would enter an accelerated phase after the first 10^{-43} seconds after the Big Bang. It turns out that in this limit, the scale factor increases exponentially with time. Thus the density of ordinary matter would quickly drop to a tiny value; in a few seconds, there would be less than one atom in the observable universe. No structure would have formed in such a universe.

Indeed, the inflation that is believed to have happened early in the history of our universe is a form of exponential expansion. But unlike the scenario we just outlined, inflation occurs at a density well below the Planck density. Moreover, it does not go on forever, but ends after a tiny fraction of a second, dumping its energy in the form of elementary particles, and beginning a hot Big Bang expansion phase of the universe.

122. Not a blunder after all?

Einstein's biggest blunder was to invoke the cosmological constant to give a static universe. Without it, Einstein's field equations of general relativity predict a universe that is either expanding or collapsing. Had he had more faith in his equations, Einstein could have predicted that the universe was not static, more than a decade before Hubble's discovery of the expansion of the universe in 1929. Moreover, including the cosmological constant to counteract the tendency of the matter of the universe to attract leads to an unstable equilibrium; the slightest nudge to the delicate balance between the cosmological constant and the self-gravity of the universe would push it to collapse or to infinite expansion.

We now invoke the cosmological constant, not to make the universe static, but to explain its observed acceleration. Einstein would be happy, because the cosmological constant idea actually becomes useful to explain an observed phenomenon of the universe. Guth's theory of inflation also invokes a period when the expansion of the universe was accelerating (albeit an expansion *much* faster than what we observe today). Therefore, he would be happy to see that physical effects do exist that can cause the universe to accelerate, just as he predicted would happen in the early universe.

123. The Big Bang

Galaxies are observed to be redshifted; their spectra are shifted systematically to the red, which we interpret as due to the expansion of space between these galaxies and us. That expansion is uniform, which is inferred from the Hubble Law, a proportionality between redshift and distance. Given this uniform expansion, we can infer a time, given roughly by the inverse of the Hubble constant H_0, when all the galaxies were on top of one another.

We term that time the Big Bang: a time of extremely high density and temperature. The inferred age of the universe is about 14 billion years, which is comfortably larger than the age of the solar system (from radioactive dating) and larger than the ages of the oldest stars (from measurements, e.g., of the Hertzsprung-Russell diagrams of globular clusters).

The Big Bang naturally explains the relative abundance of hydrogen and helium, which were formed in the first minutes after the Big Bang, as the universe expanded and cooled. In particular, the theory predicts that hydrogen should represent about 75% by mass of the atoms in the universe, with helium being 25%. The heavier elements were not formed at this time, but rather are forged in the interiors of stars. The universe is observed to have similar properties when averaged over very large scales (tens of millions of parsecs), consistent with the Big Bang idea.

The initial hot universe glowed brightly, emitting blackbody radiation. We see that remnant radiation in the form of the cosmic microwave background, light that has been traveling to us since the universe became cool enough (about 3,000 Kelvin) for neutral hydrogen atoms to form. In addition to predicting the CMB (which was discovered in 1965, winning its discoverers a Nobel Prize), the Big Bang correctly predicts

- The temperature of the CMB.
- That the CMB is isotropic, uniform, and is coming to us from all directions.
- That the spectrum of the CMB is accurately described by Planck's blackbody formula.
- That the fluctuations in the CMB have a "power spectrum" (quantifying the amplitude of these fluctuations as a function of the scale on which they are measured), in beautiful agreement with model predictions, including the amount of ordinary matter, dark matter, and dark energy observed in the universe.

124. Getting to Mars

Draw a picture of an elliptical orbit around the Sun, and draw a line along the major axis, which passes through the Sun at one focus. That line intersects the ellipse at two points, the perihelion (the closest point of the orbit), corresponding here to 1 AU, and the aphelion (the farthest point of the orbit), corresponding to 1.524 AU. Thus the major axis is the sum of the two, 2.524 AU, and the semi-major axis of the orbit is half that, 1.262 AU.

We are asked for the time to travel from the perihelion (Earth) to the aphelion (Mars) of the orbit. The major axis, which connects these two points, splits the ellipse into two halves of equal area. Kepler's second law tells us that equal areas are swept out in equal times, which means that the time to go from perihelion to aphelion is exactly half the time to make one full period.

But Kepler's third law tells us how to calculate the period, given the semi-major axis. This is an orbit around the Sun, so the simple form of Kepler's third law holds:

$$(\text{Period in years})^2 = (\text{Semi-major axis in AU})^3.$$

Plugging in 1.262 AU on the right-hand side and solving for P with a calculator gives us $P = 1.418$ years. Half that is 0.709 years, or about 8.5 months.

Note that this orbit appears in the book *The Martian*, by Andy Weir. Before the Ares 3 rocket carrying Mark Watney and the rest of the crew travels to Mars, a set of presupply probes are sent on the Hohmann orbit from Earth to Mars to deliver the supplies that the crew will need for their mission.

125. Interstellar travel: Solar sails

125.a The brightness b of the Sun, as seen from the solar sail, is given by the inverse square law:

$$b = \frac{L}{4\pi d^2},$$

where L is the Sun's luminosity, and d is its distance. This brightness represents the amount of energy per unit time, per unit area, hitting the sail. The energy per unit time impinging on the sail is that brightness times the surface area of the sail A, and thus the force on the sail, using the formula above, is

$$\text{Force} = \frac{2bA}{c} = \frac{LA}{2\pi d^2 c},$$

where c is the speed of light. We need to divide that force by the mass of the sail m to get an acceleration a:

$$a = \frac{LA}{2\pi d^2 m c}.$$

Now, we can think of the sail as a very thin sheet of area A and thickness l, and therefore volume Al. Its mass m is that volume times its density ρ. Plugging that mass into our expression for the acceleration gives

$$a = \frac{L}{2\pi d^2 l c \rho}.$$

The area of the sail has dropped out of the equation! Thus the acceleration does not depend on the area.

125.b We want the mass in the sail to equal that of the payload, 10^5 kilograms, or 10^8 grams. As we just determined, the mass of the sail is its surface area A times its thickness l times its density ρ. Solving for the area, we find

$$A = \frac{m}{l\rho} = \frac{10^8\,\text{g}}{10^{-2}\,\text{cm} \times 3\,\text{g/cm}^3} \approx 3 \times 10^9\,\text{cm}^2.$$

Note that we are working consistently in units of centimeters and grams. We are told that the sail is circular, and thus has an area $A = \pi r^2$. Solving for radius, we get

$$r = \sqrt{\frac{A}{\pi}} = \sqrt{\frac{3 \times 10^9\,\text{cm}^2}{\pi}} \approx 3 \times 10^4\,\text{cm}.$$

To do this arithmetic, take $\pi \approx 3$, and $\sqrt{10^9} = \sqrt{10^8} \times \sqrt{10} \approx 10^4 \times 3$. Thus the radius of the sail is about 300 meters; the sail is roughly 2,000 feet across. This is about twice the diameter of the Arecibo radio telescope in Puerto Rico.

125.c Here we will plug in numbers for our expression for the acceleration from part **a**:

$$a = \frac{L}{4\pi d^2 l c \rho}.$$

We've divided by another factor of 2, because in part **b**, we doubled the mass of the payload. Plugging in numbers for the luminosity of the Sun L (we'll use MKS units consistently), the distance to the Sun (1 AU), and the thickness of the sail, we find

$$a = \frac{3.8 \times 10^{26}\,\text{Joules/sec}}{4\pi(1.5 \times 10^{11}\,\text{m})^2 \times 10^{-4}\,\text{m} \times 3 \times 10^8\,\text{m/sec} \times 3000\,\text{kg/m}^3}.$$

First, let's do the units. A Joule is equivalent to $\text{kg}\,\text{m}^2/\text{sec}^2$, so the units are:

$$\frac{\text{kg}\,\text{m}^2/\text{sec}^3}{\text{m}^2\,\text{m}\,\text{m}/\text{sec}\,\text{kg}/\text{m}^3} = \frac{\text{kg}\,\text{m}^2/\text{sec}^3}{\text{kg}\,\text{m}/\text{sec}} = \frac{\text{m}}{\text{sec}^2},$$

which are the appropriate units for acceleration. Good! Now let's do the numbers. OK, without a calculator, we will set $3.8 = 4$, and cancel those terms out; $(1.5)^2 \approx 2$; similarly, $\pi \times 3 \approx 10$. Putting this all in, we find a numerical value of:

$$a = \frac{10^{26-22+4-1-8-3}}{6}\,\frac{\text{m}}{\text{sec}^2} = \frac{10^{-4}}{6}\,\frac{\text{m}}{\text{sec}^2} = 1.7 \times 10^{-5}\,\frac{\text{m}}{\text{sec}^2}.$$

That's a very small acceleration. How long would it take to get to a speed of 30 km/s = 30,000 m/s? It is simply the speed divided by the acceleration, or about 2×10^9 seconds, or about 70 years. That is an inefficient way to accelerate a spacecraft! And making the solar sail larger doesn't help us; the acceleration, as we've seen, is independent of the size of the sail. It gets worse: as the spacecraft moves away from the Sun, the intensity of sunlight decreases, dropping the acceleration even lower.

People are looking at another way to use radiation pressure to accelerate a sail, namely to use intense lasers aimed at ultralight spacecraft. Here, rather than the isotropic light from the Sun, the laser concentrates its light onto the sail. This initiative is called the Breakthrough Starshot; look it up on the web!

126. Copernican arguments

126.a Because you know nothing else, you are learning about this CEO at a random point in the full (past and future) history of her time in office. With 95% confidence, the current moment is not within the first 2.5% (1/40) of this period, nor is it within the last 2.5%. Thus at the 95% confidence level, the shortest future longevity is 1/39 its current longevity, or 1 week, and the longest future longevity is 39 times its current longevity, or roughly 1,500 weeks, or about 30 years.

126.b This is an exercise in multiplying and dividing by a factor of 39. Thus the future longevity is at least 1,000 years and is less than $39 \times 39,000 \approx 1.5 \times 10^6$ years (with 95% confidence).

126.c If you are a random person picked out of all of humanity (and it is that randomness that is the heart of the Copernican Principle), you will most likely be associated with the largest concentration of humanity, because that is where all the people are! Thus you will most likely be born in a century in which there is a large concentration of people (i.e., one with a population above the median). Simply put, most people are born in centuries with population above that of a median century.

127. Copernicus in action

127.a The Copernican model is much simpler conceptually than is the geocentric view. It explains the retrograde motions of planets in a natural way and explains the fact that the orbital periods of the planets increase the farther they are from the Sun. Direct proof that the Earth goes around the Sun includes:

- Observation of parallax of stars. Earth orbits around the Sun, and thus our perspective of the positions of stars is continually changing. While the ancient Greeks considered the effects of parallax, it is a small effect and was first observed only in the 1800s.

- Observation of phases of Venus. Venus goes through phases like the Moon does, as the face illuminated by the Sun is oriented in different ways relative to our line of sight. This effect, observable by Galileo with his primitive telescope, is a natural consequence of the Copernican model and was not explained by the Ptolemaic model.
- The parallel with the Jovian moon system, in which a large body (the planet Jupiter) is orbited by many smaller bodies (its moons). This is not a proof per se, but Galileo's discovery of moons orbiting around Jupiter drew a direct parallel with the orbits of planets around the Sun.
- The discovery of more than 3,000 planets orbiting other stars, again a parallel to our own Sun-centered planetary system.
- Observations of stellar aberration, whereby the positions of stars change systematically through the year as the direction of our motion around the Sun changes. This is a systematic shift of 20 seconds of arc, independent of the distance to the star and independent of the parallax effect.
- Observations of the Doppler shift of the spectra of stars, which change systematically through the year as Earth orbits the Sun. This is because the relative motion of Earth and the star is constantly changing.

All these discoveries were made after Copernicus died, and so he did not have direct proof that his model was correct.

127.b Early attempts by Jacobus Kapteyn and others to model the distribution of stars in the Milky Way indicated that the Sun was in the center of the Milky Way. However, they were unaware of the effect of interstellar dust, which hides our view of the Galactic center. The distribution of globular clusters is not centered on us, but rather on the center of the Galaxy, and many globular clusters lie above and below the plane of the Milky Way disk, where most of the dust is concentrated. Thus one can obtain a largely unobstructed view of their distribution. Harlow Shapley was the first to map the distribution of globular clusters, and he showed that they are in a roughly spherical distribution, centered about 25,000 light-years from the Sun. The Sun itself is embedded in a spiral arm.

127.c We observe via the phenomenon of redshift that nearly all galaxies are moving away from us, so it appears that we are at the center of the expansion of the universe. However, in uniform expansion, *every* point is moving away from any other point, so that aliens living on any galaxy also could conclude that their galaxy lies at the center of the universe. In fact, as far as we know, the expanding universe has no edge and is infinite in extent, and therefore there is no center to define relative to it. Of course, we do not know that the universe is infinite in extent; that represents an enormous (indeed, infinite!) extrapolation from the finite chunk of the universe that we do directly observe.

127.d The Copernican Principle suggests that the present time is randomly chosen in the interval of the lifetime of any process (e.g., the existence of our species, the Berlin Wall, even a Broadway play) that we might choose to consider. This allows us to put statistical bounds on how much longer the process will continue, often put in terms of a 95% confidence interval. Of course, there are cases in which the present time may not be random: for example, you could not predict the future longevity of a marriage at a wedding using these arguments, as you have been invited to see the marriage at a special time — its beginning.

128. Quick questions for our future in the universe

128.a FALSE. A unified theory of general relativity and quantum mechanics continues to elude us. Our ignorance of this question is one of the reasons we cannot give a good answer to what happens at the center of a black hole, or what happened before the Big Bang.

It is worth mentioning that quantum mechanics and special relativity are theories that do fit together quite nicely; this was first done by Paul Dirac and others in the late 1920s and early 1930s, and then developed into the full theory of quantum electrodynamics in the late 1940s by Richard Feynman and others.

128.b FALSE. The Copernican Principle states that if we choose a random moment in the lifespan of anything, we can use the laws of probability to predict a range for its future longevity. The human spaceflight program is only about 56 years old. Its future longevity is likely to be within a factor of 39 of that, if the present time is not special.

128.c The evaporation of supermassive black holes. Even the least massive stars will burn out after roughly 100 trillion (10^{14}) years. But the supermassive black holes found in the centers of galaxies will decay very slowly by Hawking radiation, taking more than 10^{94} years to evaporate completely.

128.d The fastest ticking clock you can imagine is a light clock, in which a light beam (which of course is the fastest thing possible) bounces back and forth over a short distance. The shorter the light clock is, the faster it ticks and the shorter the wavelength of the light beam has to be to fit into that clock. A shorter wavelength photon has a higher energy associated with it. That energy is equivalent to a mass via $E = mc^2$. Thus, the shorter the clock is, the greater is the mass associated with the photon. When the size of the clock becomes smaller than the Schwarzschild radius corresponding to its mass, the clock will collapse into a black hole; that is, the clock cannot exist. Therefore, there is a minimum size, and therefore tick rate, for the clock, and thus a minimum time it can measure.

129. Directed panspermia

129.a The mass is the volume times the density:

$$M = \frac{4}{3}\pi r^3 \rho \approx 4 \times (0.006\,\text{cm})^3 \times 1\frac{\text{g}}{\text{cm}^3} \approx 10^{-6}\,\text{grams}.$$

A human weighs perhaps 80 kilograms, a factor of 8×10^{10} larger. It is much easier to send the egg cell than the human to space! Moreover, we already have the technology to freeze eggs and revive them, while humans on a long space journey need full life support on the journey.

129.b This is straightforward: at 1% the speed of light, it takes 100 years to travel a light-year, so 40 light-years takes 4,000 years. The acceleration time at the beginning and the deceleration time at the end are negligible in comparison.

129.c With the 4,000-year travel time, and 10,000 years for a technologically advanced civilization to grow on each planet, each generation of galactic colonization takes 14,000 years. Our Milky Way is 100,000 light-years across, and we are 25,000 light-years from the center. So we have to go at most a distance of 75,000 light-years. If the colonization is efficient, then some rockets will always be heading radially outward: 75,000 light-years/40 light-years per step is roughly 2,000 steps of colonization. With each step taking 14,000 years, the probes would reach the farthest parts of the Milky Way in 28 million years. In doing so, would they land on all the habitable planets? The answer is yes; if each civilization puts out six probes, the total number of probes is $6^{2,000}$, vastly larger than the number of habitable planets. That is, by the time the probes reach the far side of the Milky Way, they will have visited every habitable planet, perhaps many times.

Note that 28 million years is much shorter than the age of the Milky Way. It is even shorter than the time since the demise of the dinosaurs, 65 million years ago. Thus galactic colonization can be relatively quick. But it is a long time compared to the evolutionary timescale of *Homo sapiens*. Our distant descendants who colonize the far side of the Milky Way 28 million years from now probably will not look anything like us.

129.d As described in chapter 24 of *Welcome to the Universe*, the answer is quite simple, and chilling. The Copernican Principle states that we are mostly likely to be in a non-special place among all places for intelligent observers to be. If the scenario we have painted above is true, whereby intelligent species typically go on to colonize the Galaxy, almost all intelligent observers are space colonists. The fact that we are living on the original home planet of our species would make us unusual, and thus special. Alternatively, our situation is *not* special, and most intelligent species are on their home planet. That is, space colonization of an entire galaxy must be rare.

Enrico Fermi helped develop atomic weapons. He realized, no doubt, that while atomic energy would make interstellar travel and colonization possible, it also increased the chance that civilizations could destroy themselves before getting off their home worlds. So perhaps the fact that we haven't been colonized by extraterrestrials is a sign that civilizations tend to destroy themselves before entering a space colonization phase.

129.e The moon is about 400,000 kilometers away. There are about 10^{13} kilometers in a light-year, so the ratio is

$$\frac{40\,\mathrm{ly} \times 10^{13}\,\mathrm{km/ly}}{4 \times 10^5\,\mathrm{km}} = 10^9.$$

The nearest habitable exoplanets are a billion times farther away than the Moon.

129.f One percent the speed of light is 3,000 kilometers per second. There are 3,600 seconds in an hour, and 1.6 kilometers in a mile. So the ratio of speeds is

$$\frac{3,000\,\mathrm{km/sec}}{25,000\,\mathrm{miles/hr} \times 1.6\,\mathrm{km/sec} \times 1\,\mathrm{hr}/3,600\,\mathrm{sec}} \approx 250.$$

We will have to develop spacecraft that can travel 250 times faster than our current technology has allowed humans to travel.

129.g It would take approximately $400,000/120 = 3,000$ flights of the Saturn V to deliver the raw material to build the Orion spacecraft. For comparison, only 13 Saturn Vs were ever launched.

129.h In 2017, the human spaceflight program is 56 years old. We calculated that the colonization of the Galaxy would take 28 million years, a factor of 500,000 times larger!